INDUSTRIAL
SECURITY

INDUSTRIAL SECURITY
Managing Security in the 21st Century

DAVID L. RUSSELL, PE
President
Global Environmental Operations, Inc.

PIETER C. ARLOW
Lieutenant Colonel
South African National Defense Force

For general information on our other products and services or for technical support, please contact our
Customer Care Department within the United States at (800) 762-2974, outside the United States at
(317) 572-3993 or fax (317) 572-4002.

Wiley also publishes its books in a variety of electronic formats. Some content that appears in print may
not be available in electronic formats. For more information about Wiley products, visit our web site at
www.wiley.com.

Library of Congress Cataloging-in-Publication Data:

Russell, David L., 1942–
  Industrial security : managing security in the 21st century / David L. Russell, Pieter Arlow.
    pages   cm
  Includes bibliographical references and index.
  ISBN 978-1-118-19463-8 (hardback)
1. Industries–Security measures.   2. Industrial safety.   3. Risk management.   4. Security systems.
5. Terrorism–Prevention.   I. Arlow, Pieter.   II. Title.
  HD61.5.R87 2015
  658.4′73–dc23
                                                                                          2014043896

Set in 10/12pt Times LT Std by SPi Publisher Services, Pondicherry, India

Printed in the United States of America

10 9 8 7 6 5 4 3 2 1

1 2015

For my girls and their girls:
Laura, Jennifer
Edda, Zola, and Miriam
You are all special ladies, and this is for you.
Thanks for being yourselves.
Dave Russell

"In humble submission to my Lord and Savior God,
and dedicated to my children,
Jean-Pierre, Andrich, and Landi,
who are my all here on earth"
Pieter Arlow

CONTENTS

CHAPTER 6   *BASICS OF CYBER SECURITY*   93

CHAPTER 7   *SCENARIO PLANNING AND ANALYSES*   109

CHAPTER 1

INTRODUCTION TO SECURITY RISK ASSESSMENT AND MANAGEMENT

INTRODUCTION

This course was developed out of a training outline and the course Col. Arlow and I taught together in Manama, Bahrain. Pieter's background is South African Defense Force, and he was responsible for the security of the World Cup in 2011. Dave's background is civilian, industrial chemical, and environmental consulting. Together, we believe that this book will provide a different and practical approach that combines security theory with practice. We hope that it is not just another book that is put on the shelf and used occasionally, but read and considered, and one where our suggestions are put into place.

Security is not just one group's business; it is everybody's business. The combination of security, safety, and environmental protection are critical to the operation of a modern-day chemical or industrial plant. Despite the heightened focus on security by the US Department of Homeland Security and Transportation Security Administration, in many instances, it amounts to little more than a theater of the absurd because the United States is only marginally more secure and it is more a chance of luck than of their expensive, large, and restrictive efforts to increase travel security in particular and homeland security in general. Paperwork does little to provide security.

BUSINESS DEFINITION

The business definition of security is quite straight forward. Webster's Dictionary provides us with the basis for security: "freedom danger, risk of loss, and trustworthy and dependable." That is a very good start. The definition of security crosses a number of lines in the modern industrial plant and has many different definitions. Plant security can be anything from the guard force who keeps out the unwanted intruders to the executive protection service and to the corporate watchdog that looks after the financial and corporate affairs of the plant or the corporation to make sure that there is no theft or leaking of secrets at the highest level of the company.

Industrial Security: Managing Security in the 21st Century, First Edition. David L. Russell and Pieter C. Arlow.
© 2015 John Wiley & Sons, Inc. Published 2015 by John Wiley & Sons, Inc.

With the advent of the Internet and the digital age, the job of security has been made, if anything, tougher because of the ease of communications and the proliferation of digital devices and the Internet. The communication is much easier, but then so is the ability to penetrate networks and obtain information or compromise security systems in a variety of ways. One has to look no further than the Stuxnet virus and how it delayed the development of the Iranian atomic program by attacking the centrifuges needed to refine the uranium. The success of the virus/worm delayed the development by up to 2 years.

SECURITY VERSUS RISK

In order to get a better working definition of security, we should also have a working definition of risk. Risk is the chance of loss or injury. In a situation that includes favorable and unfavorable events, risk is the probability of an unfavorable event or outcome. We measure risk by examining the certainty that a particular bad outcome or outcomes will occur.

Risk comes in many forms. There is financial risk, enterprise risk, risk of self-organized criticality (failure),[1,2] risk of injury, internal risk (theft, fire, economic loss, etc.), industrial/jurisdictional risk, operational risk, and several other types of often unforeseen and uncontrollable events that create damage. Within the various operations of a corporation, many of these have specific departments to address those risks. For example, safety, health, and environmental departments address specific risks for worker safety and environmental contamination; the IT security department manages risk for intellectual- and computer-related data. We are more concerned with the risks associated with external events such as terrorism, earthquakes, tornadoes, fire, etc. These are external risks. Internal risks might include sabotage and plant accidents resulting in fire, spills, explosion, etc.[3]

Within the scope of plant security, one is primarily concerned with events that are external to or imposed upon the plant, natural occurrences, and man-made occurrences, some of which are preventable and others not. Our working definition will include such elements as terrorism, external attacks, naturally occurring events such as tornadoes and hurricanes, and some limited scenarios for sabotage. Events such as spills, fire, and accidents may be equally unpredictable, but they are often addressable by proper design of facilities, installation of engineering controls, and management of personnel through procedures and training. Logically, we must also look into some of the process control and operational functions as a modern plant uses a variety of computer and wired and wireless control systems that are often open to sabotage or external influences.

FRAMEWORK FOR RISK MANAGEMENT

The basic framework for risk management is a cost-associated function where the general sequence starts with identification of the assets at risk, evaluation of the likelihood of their occurrence, development of a cost and a probability associated

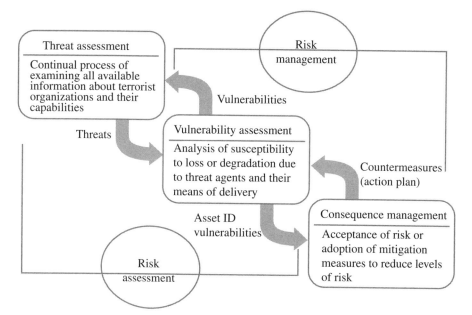

Figure 1.1   Outline of risk management actions.

with the occurrence of an attack or an event, and estimation of the costs to reduce the risk to manageable levels. This is a cyclic process, illustrated by Figures 1.1 and 1.2.

We measure and estimate the cost of a particular event occurring so that we can provide a financial plan for the plant or facility. We develop scenarios and the cost of those occurrences. For example, if we assume an attack by a hostile force, we try to estimate the damage and costs associated with that attack. We may create several scenarios and the associated costs. Things like standoff weapons such as a grenade launcher, a rocket, or a bazooka might have a damage level (cost) of C1 for the first scenario, C2 for the second scenario, etc. C1 might be for a mortar. C2 might be for a car bomb. The objective is to make these scenarios as realistic as possible when one views the likelihood of the attack.[4]

An attack can be any unplanned event and is subject to wide interpretation. Natural meteorological events can be an attack. So can an intruder into the plant. Terrorism is an attack, but then so is a civil unrest. Sabotage is a type of attack, but it is special and separate because it is imposed internally rather than from outside. However, a good risk management plan may want to consider sabotage as an element of a response plan.

Once we have a range of costs and scenarios, we can begin to determine the risk based on the probability of the events. This is often the most difficult and controversial part of the exercise because different assumptions on the likelihood of the event can produce dramatically different outcomes and costs. This is also complicated by the prospect of expenditures for increasing security and estimates as to how much specific improvements will reduce risk.

Just because a plant has *not* had an electronic intrusion (which they know of) does not mean that one will not happen tomorrow. Similarly, adverse weather events

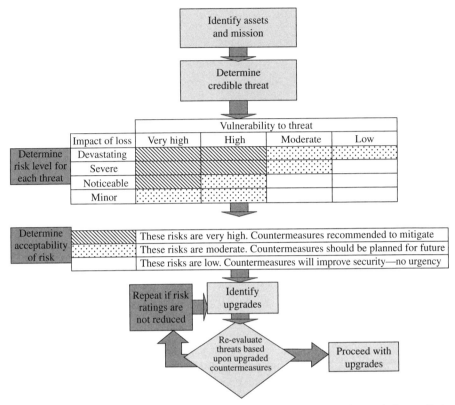

Figure 1.2   A second view of the risk analysis process. The risk analysis matrix is usually in color. Red indicates high risk, yellow indicates moderate risk, and green indicates lower levels of risk, but we have chosen to use stripes, dots, and white spaces to highlight the risk levels, respectively.

may have a record going back 30 years or more with no incidents, but that does not prove anything except that nothing has happened in that time period. History is often a very poor predictor of future events, and one needs to be careful about piling assumptions upon assumptions when and where events occur.[5] The concept of a "once in 100-year storm," popular in flood prediction and rainfall frequency analysis and other similar events, does not mean anything, except that the event was not expected with high frequency. Two of those events could occur back to back in subsequent days.[6]

In some cases, the risk assessment is relatively easy with probabilities in the percentile ranges $P=1\%$ $(P=10^{-2})$, while in many other cases, the probability of an event is on the order of 0.0001% $(P=10^{-6})$ or even less. When estimated costs and damages are high, in the millions of dollars, we have a challenge multiplying a very small probability by a very big cost. Added to this is the idea that costs are ever increasing, and the range of uncertainties is dependent upon a partial or limited database.

Fundamental to the understanding of risk are the concepts of vulnerabilities, assets, and threats. Those three components come together to form the basis for risk.

Assets are the physical structures, the data, the production, the inventory, and almost anything that has a value. Vulnerabilities are the possible methods of degrading or devaluing the assets. It is often helpful to think of vulnerabilities as the means that threats can accomplish the damage. Threats are the possible events that acting through the vulnerabilities can degrade or destroy the assets. The conjunction of all three is the risk. A word picture might help explain the concept.

A threat could be a terrorist attack by mortar or grenade or car bomb, or infiltration, or sabotage. The vulnerability might be that the main processing reactor at the facility would be damaged and that would lead to an explosion that destroyed the plant and created a fire in the storage areas, destroying them as well. The assets are the reactor, the plant, the storage areas, the inventory, and the data and might include the financial losses due to loss of revenue or accounts receivable from lost production. The assets would potentially be in the millions of dollars, but with careful planning and engineering controls, the assets could be separated to reduce the vulnerability on the scenario:

Risk = Threat × Vulnerability × Assets *and is expressed in monetary terms*

Or to express risk in another way:

Risk = Threat × Vulnerability × current or replacement cost of asset

The cost of an asset depends upon the accounting method employed and the tax structure and other variables. Generally, replacement cost for an asset needs to be updated every few years. The discussion in the following addresses some of this in very general terms.

If the threat is low and expressed in annual terms, the risk may be a few thousand dollars per year or may be diminishingly small depending upon the statistical basis employed to calculate the likelihood or probability of the threat. As we go through this book, we will try to address some of the concerns and attempt to illustrate methods to reduce the uncertainties using accepted techniques and statistical methods.

"Traditional" risk assessment programs exist to identify hazards arising from work activities to ensure suitable risk control measures are in place. However, incidents continue to happen, either as a result of inadequate risk assessments or failures in the necessary risk control measures.[7]

VALUE AT RISK

Several of the financial companies tend to look at risk a bit differently. The concept of value at risk (**VaR**) has been defined as *"the predicted worst-case loss at a specific confidence level (e.g., 95%) over a certain period of time (e.g., 1 day)."*[8] This model is being used by organizations such as Chase Bank where they take a daily snapshot of their international trading positions to determine their exposure.

The components of value can include such items as earnings, market, projected revenue, cash flow, and asset value: in short, everything. With older facilities, which may have been fully or partially depreciated, these items may be of substantially greater value than the facility itself. It should also be noted that the VaR needs to be benchmarked against a known quantity. The VaR could be actual or virtual, and may include project sales growth against a baseline or something else. The financial management of the corporation needs to be involved in deciding what is the VaR.

For example, if an attack destroys the manufacturing plant causing lost production for the principal product, the VaR might include the replacement cost of the facility, plus the value of the lost market position (sales and revenue) and lost contracts. The inclusion of these other elements in the VaR will inflate the apparent replacement costs and could conceivably cause the management of the facility and corporate management to acknowledge the value of the facility in different and perhaps improved terms.

CALCULATION OF RISK

There are various methods of calculating the probable risk. Depending upon the accounting and valuation method employed, the risk manager can use linear or nonlinear valuation methods. The methods most commonly used include techniques such as Monte Carlo simulation, parametric simulation, and historical simulation. Monte Carlo methods involved application of statistical parameters and are substantially computer intensive. Parametric and historical simulations use a combination of formulas and may involve case histories for individual cases. In the case of a plant facility, cited earlier, the valuation may require a combination of methods such as Monte Carlo methods for market risk and parametric and historical simulation methods for physical asset risks.[9]

RISK ASSESSMENT VERSUS RISK MANAGEMENT

Risk assessment and risk management are two different things. The former involves a worst-case scenario, perhaps tied to financial programming and projections, while the latter involves preparing action plans, implementing and measuring performance, and proscribing actions and objectives to minimize damage or losses. These management plans can be proactive, based on risk assessments; active, based on safety audits and site inspection; and reactive, based on incident investigation and analysis.

The selection of a particular achievable risk evaluation level is somewhat arbitrary by the plant, but note that it does tie to reality over time. A risk confidence level of 95% would indicate that the company could sustain significant losses once in every 20 days or so. While a 99% confidence interval would indicate a significant loss once every quarter. Obviously, these loss rates are unsustainable when it comes to the physical facility. The projections are more for financial risks and market risks rather than physical risks. Sustainable physical risk rates are on the order of 0.0027% (one loss in 10 years or less), and many facilities throughout the world sustain a physical risk of 0.000059% (one major loss in 30 years) or less. So a combination of loss rates and factors must be used to make an accurate calculation.

Many risks, especially those to the physical plant, are considered insurable. However, many are not. One good example of an uninsurable major risk can be found considering Superfund and CERCLA[10] Litigation. The literature and the case law are rife with cases where the insurance company had to pay for cleanup of sites contaminated by a company, and many of the insurance companies have demanded pollution riders on their policies or have denied claims for damages and cost recovery from past operations. The claims are frequently made based on real or alleged damages to local populations, health effects, and diminished values for property.[11] A number of these claims, however, are based on continuing practices rather than a specific past incident.[12]

At this point, it is also good to consider something else from the financial services industry, *stress testing*. In the realm of security, the stress test has a physical form. The military uses **red teams**, groups of individuals who are routinely cut loose from the plant structure with the specific instruction to attempt to penetrate the plant security and organize attempted security breaches and incidents. This can go to the point of planting a fake bomb, penetrating secure areas, spoofing software, and introducing harmless viruses into the operating systems of the plant. These red team activities are limited only by the ingenuity of the persons on the team and the resources available, but they should be coupled with regular drills, especially for the security personnel.

For example, the fire department runs or should run regular drills where they test their response by getting out the hoses and practicing fighting real fires. At airports, the fire companies regularly have drills that use an aircraft shell and douse it in fuel and then practice putting it out. But, there are a number of types of drills that can test the plant security and that may be appropriate. How often do we run spill drills? Similarly, if security is important, how ready is the security force able to respond to multiple incidents such as a fire or a spill *and an intruder*?

The literature is full of instances where refineries and other facilities with large tank storage have had spills that led to fires and explosions in the tank farms.[13] The point is that industry has regular firefighting drills, but when do they have security and other disaster drills? These are stress tests of the system, and the answer is, unfortunately, not so frequently. People stay sharp when they are challenged and regularly exercised on topics of concern, and increased awareness benefits everyone in the plant.

Note that in some areas, risk prevention may cross over into activities normally considered as the province of plant safety, and vice versa. If an employee is injured on the job or cannot perform his/her function, it does represent a risk to the plant finances. Similarly, the risk of employee theft or asset diversion or sabotage is also a risk. The principal difference between these and some of the previous risk factors mentioned earlier is the idea of preventable risk versus nonpreventable risk.[14] Preventable risk, such as employee risk, is often covered by safety training, procedures, and equipment. Theft, diversion of assets, and financial misappropriation are often covered by corporate security, and in the modern society, the operation and security of the plant's computers and data are protected by a special function within the information technology department. But, plant security needs a place at the "table" whenever the plant is expanded or when there are major changes to the process equipment to insure that the process is secure from outside intrusion.

Risk assessment of technological processes (chemical, petroleum, power plants, and electromechanical systems) is a complex process that requires enumeration of all possible failure modes, their probability of occurrence, and their consequences. This risk is managed through thorough analysis and technical review and playing "what if" analyses. This type of analysis is also known as HAZOPS.

RISK MANAGEMENT PLANS

We are going to march through some of the theory around risk management and develop a scenario or two and then present risk management analysis. In the following, we will not get into Monte Carlo simulation, which is often the preferred way of performing the risk analysis, but some statistics are inevitable. A good risk management plan has to cover a lot of variables and examine a lot of options. But it starts with an assessment of assets.

The first step is to start with a replacement cost assessment of the facility and its assets. This should include a valuation of the replacement cost for all equipment and might even include the cost of obtaining new or replacement permits for equipment, including such items as air pollution studies, water pollution evaluations, etc. This by itself is going to be a major effort. The risk management department of the company or the insurance provider can provide some guidance and a lot of help.

Step 1 is to obtain or develop a cost estimate for replacement of the facility.

The cost estimate should be as recent as possible, but even if it is a few years old, a fairly accurate adjustment can be made from various cost estimating handbooks, and such sources as RS Means, cost estimation, and McGraw-Hill/ *Engineering News Record*'s construction cost index. The cost estimate generally should not be any closer than two or three significant figures. Any other level of accuracy is unwarranted. The cost estimate should be broken into as many different significant production units as existing within the plant and should also include the value of associated assets and inventory. The inventory should be broken out separately, because the value of that inventory can change more rapidly than inflation.

For example:

It is the total replacement cost for the facility that will serve as the baseline for our assets in the estimation of the risk (Tables 1.1 and 1.2). Oftentimes, the asset analysis for Unit A might look like the following if we assume that Unit A is an ammonia production facility:

Remember that   **Risk = Vulnerability × Assets × Threat**

TABLE 1.1   Cost analysis for replacement of a chemical plant

Item no.	Description	Original cost (millions)	Replacement cost (2012) (millions)
1	Unit A	11.2	22.4
2	Unit B	3.7	4.1
3	Raw materials inventory A (current $)	1.1	1.1
4	Raw materials inventory B (current $)	0.3	0.3
5	Finished inventory current values A and B	5.4	5.4
6	Associated buildings and support	3.9	12.0
	Total replacement costs		44.3

TABLE 1.2   Subasset analysis for the plant in Table 1.1

	Unit A: Ammonia production facility subunit analysis		
Item	Subunit description	Damage scenario	Cost ($ millions)
1a	Reforming furnace	Furnace destroyed	3.5
1b	Combustion chamber	Chamber destroyed	0.8
1c	Shift converter/purification system	50% damage	5.8
1d	Purification system	50% destroyed (units are in parallel)	3.0
Etc.			

TABLE 1.3   Vulnerability analysis for Unit A

Item	Vulnerability	Median estimate of damage (%)	Maximum estimated of damage (%)
1a1	Explosion in the reforming furnace	20	100
1a2	Gas leak in the feed piping	10	30
2a	Combustion chamber explosion	40	100
3a	Shift converter/purification system leak	30	55
3a1	Shift converter/purification system fire	45	75
Etc.			

The next step is to consider the vulnerability. The vulnerability is dependent upon scenarios, which to some extent depend upon the threats, but the threat matrix needs to be set aside for a little while to develop the vulnerability analysis (Table 1.3).

This type of analysis gives us a baseline for vulnerabilities. Are the numbers rough? Sure. What are other scenarios you would use to develop these loss estimates and methods? The specific vulnerabilities can become quite extensive, and when one considers improvements to the process, the installation of control systems, firefighting equipment, etc. may significantly reduce the vulnerabilities to the plant, and the decision to include existing safety and equipment improvements in the baseline case is valid.

There are a number of vulnerabilities that can lead to subunits of the plant being destroyed or damaged. These are as a part of the whole and will total up to the replacement cost of the plant. These vulnerabilities need to be paired with individual threat pairs to determine the basis for the risk.

THREAT SCENARIOS

The scenarios for development of threats are subject to a wide range of occurrences and are often highly subjective. For example, the threats might be earthquake, hurricane, lightning strike, tornado, and terrorism. Some of the threats can be ruled out or assessed as highly improbable because they are statistically insignificant. For example, the threat of a hurricane in Kansas is nil, but tornadoes and lightning strikes and perhaps even earthquakes may be statistically significant.

In the threat analysis, there are a number of significant unknowns, and we do not know what we do not know, and we have to deal with that. It is the unknown that keeps risk managers up at night worrying about the level of threat and how it will be implemented against the assets.

At the end of this chapter, we are presenting a description of a chemical plant and some ideas for improvements that need to be made to eliminate or reduce potential or existing threats. In the scenario at the end of this chapter, the plant is over 40 years old, and a couple of possible threat scenarios are utilized here as examples of the types of incidents that might be used to evaluate specific threats to a plant. The following scenarios should be used with extreme caution as they are specific to an industry and only examples of the types of incidents that might occur. In our scenario, the total list of all things that need to be done is very expensive, but that is not your problem yet. Develop and prioritize the responses from the list and add new ones of your own.

A recent lecture by Kip Hawley, on the Center for Homeland Defense and Security's website (www.chds.us) pointed to an alternative model for threat analysis— Inside Out Analysis. The scenarios were based on the analysis of the Underwear Bomber who concealed explosives in his underwear and attempted to destroy himself and the plane he was flying in on December 25, 2009. That led to an reexamination of risk thinking to consider Worst Case Scenarios (or How bad could it possibly be or get?), and then plan layers of prevention toward setting up barriers against that possibility. That type of analysis is certainly appropriate in the examples and should be considered as the ultimate for response scenarios.

For example, what would happen if due to a series of catastrophic internal failures, the plant disappeared in a massive internally caused explosion which not only resulted in the deaths of critical employees but caused secondary explosions and fires and spills which led to community contamination and evacuations? Are these scenarios likely, maybe not, or even possibly not, but then that was the type of thinking which led to Buncefield, Bophal, Chernobyl, Sevesto, and the BP Texas City Disasters. The impossible is not necessarily impossible, only highly improbable. We need to consider that type of planning in our analyses.

STATISTICS AND MATHEMATICS

An aside is pertinent here. The statistics for prediction of the frequency of certain events involve fractal mathematics, and they can get somewhat involved, as the frequency follows a power curve. The events are scalable but the actual frequency of occurrence is difficult to predict. Few, if any, incidents give a warning that they are going to occur. The exception to this is hurricanes and cyclones, where the science of forecasting has enabled reasonably accurate prediction. Tsunami forecasting may provide only a few hours of warning. Earthquakes, plant accidents, and terrorist attacks cannot be predicted with any precision. Self-organized criticality is its own problem within a plant.

Dr. Ted G. Lewis, director of the Naval School of Homeland Security, has written an excellent paper on the probable loss and the frequency of occurrence. In "Cause-and-Effect or Fooled by Randomness,"[15] he discusses self-organized criticality and relates the possibility of a successful terrorist incident to a power law

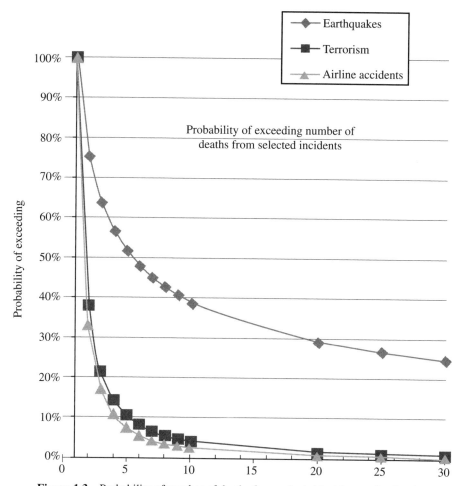

Figure 1.3   Probability of number of deaths from selected incidents, after Lewis.

function that has an expression of Probability of Exceedence $= x^{-q}$ where x is the relative severity of the incident and q is an experimentally determined exponent. Figure 1.3 is a reconstruction of some of his data from his paper. Unfortunately, the paper only has data on deaths from a terrorist incident and not financial losses.

An example calculation is that in an earthquake the probability of exceeding five deaths is approximately 50%, while the probability of exceeding five deaths from a terrorist incident or an airline accident is less than 10%.

PAIRING VULNERABILITY AND THREAT DATA

Vulnerability and threat data must be paired. There are several methods of doing that, which will be discussed in the following. The best approach for setting priorities and determining threats and vulnerability is similar to the "what if" process used by

OSHA in preparing HAZOPS. This vulnerability and threat assessment team takes a good look at the community and the surroundings and the facility. Law enforcement and perhaps military should be included in the team, along with the engineering, accounting, and security personnel from the plant. The working framework will be a group output that will identify the assets (as we have done previously), combined with an analysis of the specific vulnerabilities and the methods of attack. (Note that we are differentiating "attacks" from "incidents" because attacks are external, but incidents can include natural events and combinations of internal foul-ups, violation of safety procedures, etc.)

For an analysis of an attack scenario, one must put themselves inside the mind of a terrorist or a disgruntled employee. The attacker will not necessarily be aware of the value or criticality of the various processes in the plant but the employee will be. The intent of either of them is to cause harm or damage. The terrorist may have military-grade weapons or explosives at his/her disposal, but if plant security is reasonably good, he/she will not know where to place them for the greatest damage. The disgruntled employee will probably not have military-grade weapons or explosives at his/her disposal, but his/her knowledge of the plant is much more detailed, and he/she knows, especially within his/her department, what is most critical to the plant's operation, what is easiest to damage, and thus where his/her sabotage will do the most damage.

The questions that should be asked during an analysis of attack scenarios include such things as:

- *What are the most critical operations in the plant?*
- *Where are bottlenecks where equipment failure or sabotage or damage could shut down production?*
- *If I were an insider with intent on sabotage, how would I disable or destroy this operation?*
- *How easy is that to accomplish?*
- *If I am an outsider with limited knowledge of the plant, what is most visible?*
- *What appears to be most valuable?*
- *How would I destroy that aspect of the plant?*
- *What tools or weapons am I likely to have? How would I deploy or use them?*
- *If I have a limited knowledge of the plant, perhaps a plant aerial photo or a plant map, how would that change any of the answers to the questions above?*[16]

It is important to include the possibility of cyber attacks in the consideration of an attack scenario. The questions are slightly different and should include some of the following:

- *What type of data transmission do we use?*
- *How it is encrypted and how secure is that encryption?*
- *When we use sensors or controls or SCADA systems, are they open or encrypted?*
- *Is there cross-checking for control systems and sensors to insure that the readings are accurate and that valves and controls are operating as indicated?*
- *How easily could these systems be sabotaged or intercepted?*

- *What happens if we get a message from one of our remote sites (such as pipeline pumping station)? How quickly can we respond? What is the impact of an incident on overall production?*

- *Do we have dedicated systems for operator consoles that are not open to the Internet?*

- *Do the operators have access to the Internet and e-mail from outside sources while at their work stations?*

- *Do we have systems where anyone, including operators, can enter or extract operating data to portable electronic media? (Do the computer systems have USB ports or CD/DVD reading/writing systems in the computer?)*

If these questions do not give the plant security task force nightmares, the plant is either very secure, or the task force is asleep and uninvolved and does not realize the potential for internal or external attack. The purpose of the questions is to lead the task force to prioritize and expand a risk table similar to the one shown in the following.

SETTING PRIORITIES

The challenge of setting priorities is inexorably linked with the determination of the likelihood of the events or attacks, and the entire process is influenced by the cost of the attacks and the cost of control or mitigation measures designed to prevent the attacks or minimize the damages. There are a number of methods of setting the priorities and determining the likely annual and other costs. There are also a number of very good software packages that will assist with this effort. First, we will discuss the basic methodology and then one example of good software for helping determine priorities.

The basic risk analysis matrix is usually expressed as a table using stripes, dots, and white space, for easy identification of the levels of risk and the costs to some agreed-upon basis. The key words are "agreed-upon basis." Where there is no agreement or definition, the determination of risk is just opinion (Table 1.4).

This is a risk table, and in one form or another, it is used and modified to perform risk analysis in facilities. By adding rows and columns and performing advanced analysis on the table elements and by carefully defining vulnerability, we can develop

TABLE 1.4   Example of risk analysis by table

		Criticality rating			
	Vulnerability rating	Cost very high	Cost high	Cost moderate	Cost low
Item descriptions	Very high				
	High				
	Moderate				
	Low				

The risk analysis matrix is usually in color. Red indicates high risk, yellow indicates moderate risk, and green indicates lower levels of risk, but we have chosen to use stripes, dots, and white spaces to highlight the risk levels, respectively.

a reasonably accurate estimate of the risk and the cost to reduce it. There are a few important points to remember about this type of analysis: (i) one cannot foresee everything, (ii) much of the data are subjective and may be accurate only to orders of magnitude, (iii) performing this type of risk analysis can be a lot of work, (iv) the likelihood of an attack or the frequency of an event will probably be the most difficult element to estimate, (v) put political considerations and internal disputes aside because what is done is for the health and survivability of the entire plant, and (vi) garbage in, garbage out (GIGO)! When you calculate the numbers, get serious about it or do not perform it.

OTHER DEFINITIONS OF RISK ASSESSMENT

"Traditional" risk assessment programs exist to identify hazards arising from work activities to ensure suitable risk control measures are in place. However, incidents continue to happen, either as a result of inadequate risk assessments or failures in the necessary risk control measures.

Risk management involves preparing action plans, implementing, and measuring performance. This can be proactive, based on risk assessments; active, based on safety audits and site inspection; and reactive, based on incident investigation and analysis.

Risk assessment of technological processes (chemical and power plants, electromechanical systems) is a complex process that requires enumeration of all possible failure modes, their probability of occurrence, and their consequences.

Security risk analysis, otherwise known as risk assessment, is fundamental to the security of any organization. It is essential in ensuring that controls and expenditure are fully commensurate with the risks to which the organization is exposed. However, many conventional methods for performing security risk analysis are becoming more and more untenable in terms of usability, flexibility, and critically—in terms of what they produce for the user.

The basic elements of risk must be explored, and a security risk assessment methodology and tools must be introduced to help ensure compliance with security policies, external standards (such as ISO 17799), and legislation (such as data protection legislation).

BUSINESS DEFINITION FOR RISK ASSESSMENT

Determining the level of risk in a particular course of action is important. Risk assessments are an important tool in areas such as health and safety management and environmental management. Results of a risk assessment can be used, for example, to identify areas in which safety can be improved. Risk assessment can also be used to determine more intangible forms of risk, including economic and social risk, and can inform the scenario planning process. The amount of risk involved in a particular course of action is compared to its expected benefits to provide evidence for decision making.

BROAD DEFINITION FOR RISK ASSESSMENT

Risk assessment is the overall process of identifying all the risks to and from an activity and assessing the potential impact of each risk. The impact is measured by combining assessed and costed risk, the likelihood of an incident, and the impact of the incident. These elements are then combined to produce a single cost figure.[17]

QUANTITATIVE RISK ASSESSMENT

This approach employs two fundamental elements: the probability of an event occurring and the likely loss should it occur. Quantitative risk analysis makes use of a single figure produced from these elements. This is called the "annual loss expectancy (ALE)" or the "estimated annual cost (EAC)." This is calculated for an event by simply multiplying the potential loss by the probability. As previously discussed, it is theoretically possible to rank events in order of risk (ALE) and to make decisions based on it accordingly. The problems with this type of risk analysis are usually associated with the unreliability and inaccuracy of the data. Probability can rarely be precise and can, in some cases, promote complacency. Controls and countermeasures often tackle a number of potential events, and the events themselves are frequently interrelated, and the cost of improvements cannot be clearly calculated or assigned.

Notwithstanding the drawbacks, a number of organizations have successfully adopted quantitative risk analysis.

QUALITATIVE RISK ASSESSMENT

This is by far the most widely used approach to risk analysis. Probability data is not required and only estimated potential loss is used. If we are cynical and not willing to perform the work required to make our risk assessment quantitative, we would call this informed opinion. It is easier to perform but harder to justify, especially to the financial types in the plant environment. Most qualitative risk analysis methodologies make use of a number of interrelated elements.

Threats

These are things that can go wrong or that can "attack" the system. Examples might include fire or fraud. Threats are ever present for every system. We discussed a number of these earlier.

Vulnerabilities

These make a system more prone to attack by a threat or make an attack more likely to have some success or impact. For example, for fire, a vulnerability would be the presence of inflammable materials (e.g., paper or stored hydrocarbon liquids or even flammable gasses). Again, it is often difficult to express a quantitative or qualitative percentage of the operations that would increase the vulnerability. The easiest way of expressing the vulnerabilities is through group consensus, which leads to an agreed-upon percentage for damage.

COUNTERMEASURES FOR VULNERABILITIES

There are four types of **controls** that are critical countermeasures for vulnerabilities:

1. **Deterrent controls**—Deterrent controls reduce the likelihood of a deliberate attack. If one cannot see or locate the target, it cannot be attacked. Similarly, if the site or perimeter is so intimidating, the likelihood of attack is reduced. (Think of trying to attack a castle high on a steep and rocky hill; see photo in the following.)

2. **Preventative controls**—Preventative controls protect vulnerabilities and make an attack unsuccessful or reduce its impact. Preventive controls would consist of things like fireproof construction, double-walled piping, and SCADA systems with shutdown and alarm systems in the event of a system upset. Equipment such as safety flares, vents, and pressure relief vents would also fall into this category.

3. **Corrective controls**—Corrective controls reduce the effect of an attack. Fire sprinklers, fire brigade, blast walls, spill control diking, and relocation of certain facilities to give greater separation from hazards would all fall into this category.

4. **Detective controls**—Detective controls discover attacks and trigger preventative or corrective controls. This is the area for inspections and preventive maintenance, sensor systems, radar, television cameras, computer monitoring, and facial recognition services.

These controls are outlined in Figure 1.4: it illustrates the manner in which these controls work.

The D's of security systems

There are "**three D's**" of security: "**denial**," "**detection**," and "**deterrence**."

Another set of "**D's**" is as follows: "**detect**," "**delay**," "**defend**," and "**devalue**."

Detection is the easiest to explain. It involves identifying the attack prior to its inception. Surveillance, fencing sensors, remote detection devices, community intelligence, cooperation with the local police and military authorities, development of a local community network of informants and contacts, monitoring of the employees and their families all could contribute to the detection phase.

Denial generally involves design and construction that prevent the facility from being attacked. In this sense, a fence or barrier contributes to the denial, just as a locked door does. So does a concrete barrier system and/or bollards. In the design of denial systems such as traffic controls, it is often wise to consider the potential impact of car or truck bombs and barrier systems that will reduce or minimize the potential effects of blasts.

Deterrence is related to the presence and visibility of a force or a force projection that says to the potential attacker, "Don't even think of trying to attack this facility." It is a projection of the visible symbols and facilities of the security and

Introduction to risk analysis

These elements can be illustrated by a simple relational model:

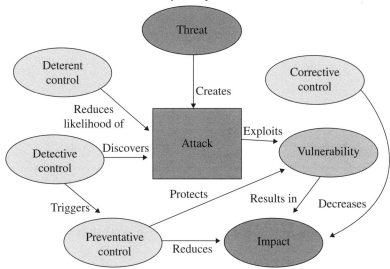

Figure 1.4 Graphic of the functioning of controls.

defense of the plant. For example, on the way into one of the large Aramco facilities in Saudi Arabia, the approach road is lined with Jersey barriers[18] to direct traffic, visible security checks, and a highly visible security presence, and after the security checkpoints, there are armed defensive positions with automatic weapons that are rapidly accessible from the security positions. The guard force is also equipped with two-way radios, side arms, and the aura of an armed presence that could repel an attacker. This can be a strong deterrent to an attack.

When it comes to the terms "detect," "delay," "defend," and "devalue," the "delay" and "devalue" terms may need a bit of clarification. Most conventional fencing is looked at as a delaying presence. However, most chain link fencing will only delay a determined intruder by less than a minute, and a locked door may only provide slightly more than a minute of delay.

A recent incident related to me by someone in the oil industry in the Middle East indicated that a major oil facility had triple layers of fencing, fairly widely separated with separate zones of influence, and military response for the middle zone. Each line of fencing was separated by several tens of meters.

Two terrorist suicide bombers in separate trucks coordinated their attack on the facility. The first one made it past the outer perimeter, but could not penetrate into the middle perimeter. He started ramming his truck laden with explosives against the fence in an effort to batter it down. The terrorist was so focused on his own activities and failures that, in frustration, he detonated his truck bomb just as the second terrorist drove up to the fence. The resulting explosion took out both trucks, but aside from a hole in the ground where the fence was and some flying debris, the facility was undamaged. In this instance, the value of delay was important both for response and for outcome.

The concept of "devalue" is akin to the concept of disguise or camouflage. What an attacker can see and if highly visible may be a target. He may know that a

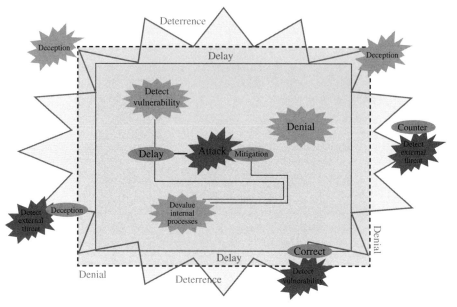

Figure 1.5   The D's of security.

pipeline exists, but if it is underground and not visible, it may not really be a target worthwhile for his consideration—similar to buildings under cover or screen, fencing that uses screening, and barriers that restrict or prevent observation. Consider that even the most dedicated of attackers using standoff weapons will want to see some of his aiming point and/or have the satisfaction of knowing that he has created some damage. It is almost like saying, "Out of sight, out of mind" (Fig. 1.5).

First, consider a logical expression for security vulnerability, which is usually expressed in terms of the probability of adversary success given attempt:

- $\Pr(S) = 1 - \Pr(\text{detect}) \cdot \Pr(\text{engage}) \cdot \Pr(\text{neutralize})$
- In words, this equation states that adversary nonsuccess (defender success) requires that the defender detect, engage (which consists of delay and response), and then neutralize the adversary (in sequence)—failure to do any one of these will result in adversary success (barring any random things outside the protector's control that might thwart the adversary's attempt).

SAMPLE THREAT SCENARIO NO. 1 (FIG. 1.6)

Background
The process   You are the chief security officer for a large chemical complex that manufactures ammonia, urea, ammonium nitrate, and urea-formaldehyde products (based on an actual chemical plant). The plant is about 40 years old and uses natural gas as a feedstock in a Haber–Bosch gas shift reactor. Natural gas is partially burned to produce CO and H_2 and purified. Then the H_2 is reacted with the leftover N_2 from

Figure 1.6   Ammonia plant complex in Ohio, United States (40-year-old picture).

the partial combustion and recombined in a recirculating loop to produce ammonia (gas). The ammonia gas is reacted with CO to make urea, and the ammonia is then burned to produce NO and reacted stepwise with oxygen to produce N_2O_4 and then with water to produce nitric acid or HNO_3. The nitric acid is then reacted with more ammonia to form NH_3NO_3 or ammonium nitrate. Ammonium nitrate is a liquid that is pumped through a shot tower where it passes through a fine screened tray to form droplets that are then dropped against a rising column of air to form spherical solids or prills. The prills are hydroscopic and will adsorb moisture from the air and melt together, so they are coated with a fine coating of wax to prevent their melting together.

The plant   The plant is a sprawling conglomeration of several separate individual manufacturing operations. The plant units are well separated by several hundred meters, and they are essentially stand-alone units. The plant has a large rail facility (10 separate spur lines from the main line) adjacent to its manufacturing and equally large storage areas (warehouses) for shipping and receiving. The plant footprint is approximately 700 acres (300 ha) and is protected by a 2 m chain link fence on three sides, and one side is along a major river, where there is a commercial boat dock for water shipments. The plant also has major storage areas for diesel oil and a major gas holder for the natural gas supply. About 1/2 of the plant area is devoted to the shipping/rail yard. Process units are separated by about 100 m. The plant uses coal as a basic source of energy and receives it by rail. Most of the shipments into and from the plant are by rail. The rail yard is separated from the main plant by an internal chain link fence, and the plant has two of its own locomotives (mules) to move the rail cars around inside the plant.

The area surrounding the plant consists of a four-lane divided highway (moderate traffic) and a two-lane access road. On two other sides of the plant, there are commercial developments and small residential developments. The plant shipping areas (tank farm) have 6–500,000 gallon insulated tanks (two for liquid ammonia, two for other chemicals, and one for diesel fuel) all diked with earthen dikes and an 80 ft diameter high-pressure ammonia storage tank. The plant dikes do not have good drainage, and manual valves are used to drain rainwater, and as a consequence, many valves are left open.

The plant computer facilities are at least 10 years old, and many of the orders arrive by fax, and confirmation is often sent out by wire and by antiquated teletype and printer systems. The plant control systems are reasonably modern, but they have an open system with distributed control systems and wireline controls for various processes.

Plant surveillance is at a minimum. There are some TV cameras but they are primarily at the plant office and the main gate. The guard force is 20 people, and most of the time, the guards are involved in the traffic management in and out of the plant. There are three guards on night shift, and the guard force office is right by the front gate where all the communications is located, including the plant switchboard and plant radio.

The plant is union and has approximately 2700 employees in a 24/7 operation. About 300 of the personnel are office workers and are on a 5-day per week schedule. It has its own fire department (small) and there is a community fire department (one truck + volunteers) in the adjacent community. There are two sizeable cities across the river, plus other emergency services including two hospitals and other commercial services available, but they are 0.5 hour away by car.

The rail yard is poorly illuminated, but given the nature of the facility, it is not considered much of a risk (no TV cameras). Most of the plant is of open (unprotected) construction, with piping and process areas open to the weather, except for control houses in each facility, and compressor and electrical areas and a break room. The plant is maintained in a reasonable manner, but there is always a persistent and strong (eye-watering) level of ammonia fumes in the compressor buildings and in the process areas. Operators must wear respirators and goggles when working in the compressor areas, and goggles are mandatory in all parts of the plant. Hard hats and long sleeve clothing are also mandatory.

In the event that an employee becomes contaminated by ammonia, the standing instructions are to head to the showers and begin undressing while in the shower. (The safety showers are heated and weather protected.) Activating the safety shower will summon the rescue personnel, and the standing instructions are to completely undress, including shoes and all clothing in the event of contamination. Safety personnel will wrap the person in protective blankets and will provide medical treatment as appropriate on the way to the infirmary.

The threat   There is general unrest and vandalism against the plant. The plant union is threatening a strike, and you have specific community individuals who are stirring up anger against the plant and trying to get the communities organized to ask the plant to shut down or move. There is additionally some threat against the idea that the plant is manufacturing "high explosives" (ammonium nitrate), and that is viewed as a potential threat to the community as a whole.

Your challenge   You are a new head of security. Given that the plant is in relatively poor shape, you have been tasked to come up with a plan to improve security. A rash of thefts and vandalism has alerted the plant management, and you have the authority and significant budge to upgrade physical security at the plant. What are your priorities?

Note: this is a composite of an older plant fabricated from work memory of five different chemical plant locations each at least 40 years old. You will also want to look up some of the chemical properties of ammonia, ammonium nitrate, and CO and familiarize yourself with the overall process before attempting to provide answers or prioritizing some of the more complex responses involving the community.

Ammonia has some toxic and irritating properties in its various forms. A quick look at a material safety data sheet (MSDS) will help clarify some of the issues. Ammonia can be both smelly and irritating to mucous membranes, including the lungs and the eyes. Given high concentrations of ammonia, inhalation can be fatal, but that happens only after eye irritation and lung irritation. In a community setting, it can be extremely annoying.

Nitric acid generates a pungent and irritating vapor. Again, look up the MSDS for data on the compound. While nitric acid is not generally a substance released to the environment either accidentally or deliberately, it is a corrosive liquid. It is not generally a fence line issue for the plant.

Ammonium nitrate is unique because it is both a fuel and an oxidizer. Ammonium nitrate is a fertilizer and a primary high explosive, and when mixed with fuel oil, it becomes ammonium nitrate fuel oil, the explosive which was detonated in the destruction of the Murrah Building in Oklahoma City, OK, on April 19, 1995. This was a "home-grown" terrorist act. For reference of the power of ammonium nitrate by itself, look up the Texas City disaster of April 16, 1947, which was the largest nonnuclear explosion in the United States (see Wikipedia on the issue for detailed description), and the West Fertilizer fire in West, Texas, on April 17, 2013. The West Fertilizer fire may be more directly applicable to the local residents' concerns about the liability from having the plant in their neighborhood, which may need reassurances that the plant safety and security is performing adequately.

A list of possible priorities   There are a huge number of problems in this scenario. Poor management and lack of attention to maintenance and community have led to vandalism, which may be internal or external to the workforce. There are problems with the community, with the regulatory community, with the work force, and with lots of other areas. The guard force is minimal given the size of the plant.

Here are some solutions and recommendations that will help improve things. We have deliberately not prioritized the order of the improvements because perusing the list will cause you to consider what you might do first. Go through the list and put some numbers on the list to focus on the things you think are important:

- Add new guards.
- Rotate shifts, and add a standby reserve.
- Dramatically and visibly increase security presence in the vicinity of the ammonium nitrate plant and its storage areas.
- Improve communications with the local police and fire and emergency services.
- New guard stations with keys for verification of guard surveillance.
- Greatly increase video surveillance.
- Duplicate communication/surveillance system at a secure location inside the plant.
- New encoded plant radios.
- Patch holes in plant fence.
- Replace dike drain valves with rising stem valves to increase visibility of closure.
- Better lighting/video surveillance in rail yard.
- Improve diking/spill control in loading areas.
- Badge all workers and install gate controls.
- All workers must wear badges at all times.
- Color-code hard hats and equipment for better visibility and identification.
- Biometric controls in sensitive areas of plant.
- Set up community contact committee for better local relations.
- Better training/retraining for guard force.
- Plant-wide meetings to explain new security and management procedures.
- New color video surveillance equipment.
- Better plant perimeter lighting.
- Institute safety and security drills with multiple events.
- Provide occasional plant lunches to various departments when safety/security goals have been met.
- Install a guard station at the point that the railroad enters the plant.
- Lock the railroad gate, install video cameras, and inspect station/shelter at the gate location.
- Coordinate the movement of rail cars with the warehouse/shipping departments and their supervisors.
- Install security on distributed control systems.
- Upgrade plant-wide internal and external communications with secure and buried fiber-optic lines and backup radio links.
- Light pathways and work areas to provide bright walkways and operator areas for night operations to increase worker inspections.

- Examine fence lighting to provide patrols with dark spaces, but illuminate and perform video surveillance of the fence line.
- Insure that foliage on the outside of the outside of the fence has been cleared to at least 20 ft from fence line.
- Install new lighting and video surveillance in the warehouse areas.
- Install fire and emergency alarms throughout the plant areas.
- Install a weather station with recording gas detectors on the fence line near the communities.
- List your other actions here:
- 1
- 2
- 3
- 4
- 5
- 6
- 7
- 8
- 9
- 10
- 11

SAMPLE THREAT SCENARIO NO. 2

Figure 1.7 is an aerial view of a large chlorine–caustic soda chemical plant. For security purposes, the location has not been identified, and the photo is approximately 40 years old. The plant is adjacent to a city in the Northeastern United States, and parts of the plant are over 100 years old.

Background

The plant produces chlorine (gas—Cl_2) and caustic soda (NaOH) through an electrolysis process. When saltwater is disassociated, it produces chlorine gas, sodium hydroxide, and hydrogen. Sodium hydroxide is a highly corrosive base, and chlorine gas is highly toxic and very corrosive as well. Chlorine gas was used in WW I as a chemical warfare agent, and it is being used by the Syrian regime against the rebels in the 2014 conflicts. The gas is dense and has a sharp odor and is yellow green in color. You are referred to the MSDS on both products for information on the toxic properties.

The plant   The plant is old and has been around since the early 1900s. Consequently, there is asbestos in almost all parts of the plant, so construction or improvements need to be carefully managed. The plant uses coal as a primary source for its steam,

Figure 1.7 Chlorine plant complex in New York, United States (40-year-old picture).

and recently, the plant has constructed a new recycling facility that will generate energy from residential and commercial garbage. The energy from waste (EFW) facility accepts truckloads of garbage and industrial wastes and sorts and classifies it to gain energy. The EFW plant has had significant startup problems including air problems, odor problems, and significant material handling problems, which resulted in poor performance, several fires, and other operating problems even though the EFW plant is highly automated.

As part of the company safety plans, the plant has a resident doctor and the infirmary is active 24 hours per day. Workers in the plant must wear proper safety equipment at all times, including a hard hat, boots or work shoes, goggles, and an escape canister. If there is a chlorine release, the escape canister reduces the concentration of gas in the area to breathable levels, giving the employees the opportunity to run away from the gas plume. There is also a gas leak alarm at the plant, and everyone is accustomed to hearing it and knowing that they are to run away from gas clouds. As shown previously, there are residential areas on two sides of the plant and associated heavy industry on a third side. The plant is bounded by a highway on the south and bisected by another and also by the railroad spur. The northern portion of the plant is used by the EFW facility. The plant has a 2 m chain link fence with three strand barbed wire on top. At night, several of the principal operating areas associated with the chlorine plant are brightly lit, but the rail yard and nonprocess areas are poorly lit to unlit.

The threat   The plant has a strong and militant union that has gone out on strike frequently. The plant employs about 280 persons in a 24/7 operation. The management and the union are often at odds, and in a recent strike, there were a number of minor acts of vandalism, sabotage, and pranks. There is a major waterway near the plant, but given the nature of the waterway, most of the chemicals are shipped out via rail through highly populated areas. The rail line separates the two halves of the plant, and there is a growing community awareness that the chemicals manufactured and their derivatives are toxic, harmful, and potentially life threatening. There is no current way of determining who is in the plant at any one time. There is also the issue of an occasional chlorine release into the community, sometimes strong enough to cause the paint on the adjacent houses to need repainting.

The local community is near or at a boiling point because of the plant, the waste problems, and other issues related to the long-term disposal of chlorinated chemicals in a local landfill. Cars of employees have been vandalized, some of the buildings have been tagged with spray paint, and plant intrusions have been recorded. So far, every incident is relatively minor. There have been threats of bombs and derailing shipments and sabotage by militant environmentalists and others opposed to the plant's presence.

The security is divided because part of the staff, approximately half, is relegated to the north gate to control traffic and security for the EFW facility (north of the tracks). There are two guard stations, but the north gate and south gate stations operate independently and do not really communicate. Trucks entering the plant are directed to the warehouse. The dispatch and inspection stations are located right on top of the plant gates.

Your challenge   You have just been made of head of security and have about 20 people on your staff. Your new position is as a result of your boss being fired for an incident involving diversion of funds in conjunction with activities in the warehouse. You are now reporting directly to the vice president of operations. He called you in with your new appointment and told you to quiet the community around the plant and improve security at the plant. Your challenge is to come up with your own list of actions and improvements to make the plant better liked in the community. Develop a list of at least 10 action items, but do not worry about the budget.

NOTES

1 See the writings of Dr. Ted G. Lewis of the Naval Postgraduate School for Homeland Defense and Security. He has an extensive model simulation of self-organized criticality on the website https://www.chds.us/?media/resources&collection=53&type=SIMULATION.

2 As an example of self-organized criticality that leads to failure, one frequently has to look no further than any one of many government assistance programs or regulatory programs. In the years that they have been in existence, the bureaucracy has taken over the mission, and now, services are being delayed or denied because of a paperwork blizzard, deliberate cover-ups that point to political agendas, and sheer incompetence. One of those agencies in the United States is the military veteran's hospital system where care is being denied to needy veterans and the paperwork has been shuffled around to make the system appear better than it is. Similar charges have been leveled against the Internal Revenue Service, the Department of Homeland Security, the Department of Interior, most of the Department of Health and Human Services, and the Environmental Protection Agency where blind adherence to rules has often trumped common sense.

3 One of the methods to reduce internal risk is HAZOPS, a critical review of plant operations and processes with a focus on analysis of potential methods of failures. For detailed requirements of HAZOPS, see the website www.OSHA.gov and search for process safety management and HAZOPS, or see http://www.osha.gov/Publications/osha3133.html.

4 Sabotage can be very serious or just annoying. In plants with strong unions, it can be a factor during a strike. Based upon personal experience, one union member boiler house worker ran a hose into the inspection port of the hopper in the coal-fired boiler cyclones immediately prior to the strike. The water interacted with the fly ash and made cleaning the collection hopper very difficult. The hoppers had to be cleaned continually in order for the cyclones to remove the fly ash. Sabotage? Sure, but more of an annoyance rather than a damage.

5 A good case in point is illustrated by the BP refinery disaster in Texas City on March 23, 2005. In that incident, as Wikipedia described, "A number of plant workers were killed when a vent stack flooded and spewed volatile chemicals in the vicinity of the plant. Hydrocarbon flow to the blow down drum and stack overwhelmed it, resulting in liquids carrying over out of the top of the stack, flowing down the stack, accumulating on the ground, and causing a vapor cloud, which was ignited by a contractor's pickup truck as the engine was left running. The report identified numerous failings in equipment, risk management, staff management, working culture at the site, maintenance and inspection, and general health and safety assessments." One of the CSB findings was that a trailer that was occupied only a small fraction of the year was located too close to the stack (source of the explosion). The trailer was occupied at the time of the explosion, resulting in over a dozen fatalities.

6 Any competent meteorologist or hydrologist will tell you that the probability of an extreme weather event is highly unpredictable and the frequency of the event is cited with respect to an anticipated average frequency. The problem is that the database from which that frequency data was developed is often less than 100 years.

7 A recent visit to a petroleum refinery in the Middle East illustrates this statement. Plant maintenance removed a pump from a bottom fed line on a large crude oil storage tank. Nobody told the operators that the pump had been removed. The tank valve was not locked out nor de-energized. "According to the plant operator, 'the valve came open by itself.'" The resulting spill was 55,000 barrels of crude oil. Only 50,000 barrels of crude was recovered due to poor design of the impoundment. The other 5000 barrels are floating above the saline groundwater and are gradually making their way toward the ocean. These types of accidents generally happen through multiple failures to follow procedures, and the after

incident report almost always indicated that the accident was preventable if procedures were followed.

 8 **Risk Management: A Practical Guide**, ©RMG Risk Management Group, 1999.
 9 Physical and historical assessment methods may involve such items as building cost data adjusted by historical inflation and building costs indices such as the *Engineering News Record* (magazine) index and such estimating guides as RS Means guides on building and equipment cost reconstruction. The DOE also has a guide on cost estimating, as do many of the other agencies, including the DOD and the GAO. Some of these guides are quite specific. McGraw-Hill Publishing Company has a software program for equipment costs as part of the *Plant Design and Economics for Chemical Engineers*, 5th ed., and there are a number of other equipment cost estimating guides on the Internet. Also, many of the large engineering and design-build construction companies also have very good equipment estimating divisions.

The accuracy of a good estimate increases with the amount of detail and the cost of preparing that estimate. The cost of a good estimate accurate is within about 10% of the actual construction cost.

The American Association of Cost Estimators has defined five categories of accuracy for cost estimates. Their most accurate estimate category is a Class 1 estimate, which has a range of −3 to +15% of construction cost, and the estimates are made from a detailed takeoff of bid and process documents, but these cost estimates are expensive and can run to 10% or more of the construction cost of the project.

10 CERCLA, the acronym for **Comprehensive Environmental Response, Compensation, and Liability Act**, authorizes EPA to respond to releases, or threatened releases, of hazardous substances that may endanger public health, welfare, or the environment. CERCLA also enables EPA to force parties responsible for environmental contamination to clean it up or to reimburse the Superfund for response or remediation costs incurred by EPA. The Superfund Amendments and Reauthorization Act (SARA) of 1986 revised various sections of CERCLA, extended the taxing authority for the Superfund, and created a free-standing law, SARA Title III, also known as the Emergency Planning and Community Right-to-Know Act (EPCRA).

11 One of the most notorious incidents is the Bhopal, India, incident where, in 1984, an alleged incident of sabotage killed over 3787 and injured 558,125 persons through a gas release of methyl isocyanate. The cause is under debate with the Indian Government claiming the cause was negligence and sloppy management, while Union Carbide (now Dow Chemical) claims that the cause was an act of deliberate sabotage.

12 Under tort law, which is often largely unwritten, many major suits for equitable relief have been brought against companies for permitting or releasing contamination into the environment and causing property damage and personal injury to residents of areas where the company has operations, either in or out of the United States. US cases of note include the Love Canal, a case about a leaking landfill; the Hinkley, CA, hexavalent chromium-contaminated groundwater incident against Pacific Gas and Electric brought to notice by the film *Erin Brockovich*; the Times Beach incident involving polychlorinated biphenyls; and the Woburn, MA, groundwater contamination (chlorinated solvents) incident, which was described in the movie *A Civil Action*. There are many, many others.

13 Two significant refinery fires and explosions that lead to fatalities include Buncefield Oil Storage Depot, Hertfordshire, United Kingdom, in December 2005, and the Gulf Oil refinery fire, Philadelphia, on August 17, 1975 (see Wikipedia for descriptions of each incident). There was also a significant refinery fire that resulted in fatalities of the firefighters in 1975. The drainage system was insufficient to remove the oil from a tank fire, and the firefighters were working in a pool of water that had oil on it. When the

fire fighting foam blanketing a pool of water with oil standing on it developed an opening, vapors were released and ignited, and eight firefighters were killed or died as a result of burns (see the article on the fire in Wikipedia).

14 A discussion with a prominent Middle East security force for a major petroleum producer indicated that right after the World Trade Center attack on September 11, 2001, the security force went into a crash priority program of reorganization because they realized that a number of the perpetrators were of Arabic nationality and might be coming after their facilities next. Prior to that, they were focused primarily inward until they realized that they had an external threat as well.

15 Dr. Ted G. Lewis is a professor of computer science and an executive director of the Center for Homeland Defense and Security at the Naval Postgraduate School. The book citation is as follows: Lewis, Ted G. "Cause-and-Effect or Fooled by Randomness?" *Homeland Security Affairs* 6, issue 1 (January 2010). http://www.hsaj.org/?article=6.1.6.

16 In a recent training course at a large chemical company in the Middle East, some elemental network analysis indicated that the location of the security command center was right next to the front gate. Any terrorist incident involving a bomb would have taken out that command center, leaving emergency services coordination in a difficult situation. The solution to that problem is to provide a backup command center with full duplicated access to everything that the primary command center has in the way of information. The information sources, however, cannot be routed through the primary center, but must be fully independent in order to be reliable.

17 Source: BNET Business Dictionary.

18 A Jersey barrier looks like an inverted letter Y. It is about 0.7 m tall, 10–15 cm at the top and about 0.5 m at the base. Effective ones are made from concrete.

CHAPTER 2

RISK ASSESSMENT BASICS

STREET CALCULUS AND PERCEIVED RISK

Risk is around us everywhere. We cannot begin to understand specific risks without an evaluation of the nature of relative risk and our perception of it. We perform a risk analysis whenever we express fear, concern, or doubt. The fear of failure, injury, death, or loss is something we experience every day, and we express fear when our "street calculus" tells us that the risk we are about to take may be unacceptable and will result in injury, death, or loss.

Similarly, we look at plant operations and fail to consider the risks that are over and above what we consider "normal" operations. Familiarity with those risks often leads us to dismiss them, saying "It will never happen here," or ignore them completely. Often, plant management wants to look only at "special" risks, which are unusual events not part of daily operations. True plant security must consider natural or routine risks as both internal and external to the plant environment.

Street Calculus

Street Calculus is the conscious and unconscious evaluation of risk. A collision, an accident, or a harmful event is a failure of risk assessment. We perform a "Street Calculus" or small risk assessment whenever we cross the street, drive through a strange neighborhood, or encounter someone on the street. There are as many different ways of assessing risk as there are people. There is no right or wrong way, but some ways are more complete than others.

Imagine yourself walking down a street at dusk when the visibility is not so good. In the gloom, you see someone approaching you, and they cross the street to come toward you. What do you do?

Do you avoid the person, turn around, and walk away quickly, or do you meet the person head on? What is your risk level before you recognize the individual? An example of a low risk might be when you get closer you recognize that the person approaching is either a police officer or a little old lady of small stature. Both are perceived as relatively low-risk persons.

Industrial Security: Managing Security in the 21st Century, First Edition. David L. Russell and Pieter C. Arlow.
© 2015 John Wiley & Sons, Inc. Published 2015 by John Wiley & Sons, Inc.

However, if you are a criminal or carrying illegal drugs in your possession, you may view the policeman as a high-risk threat rather than a person of low risk.

There are mitigating factors in this scenario. You are carrying a FedEx package you have just picked up, and as the person approaches, you determine that it is a little old lady carrying groceries. The mitigating factors change the relative risk equation from one of potential suspicion and hazard to one of nonhazard.

Now, changing the scenario slightly, you are the police officer and have just been told to be on the lookout for a young man of swarthy complexion wearing a baseball cap, sweatshirt, and jeans, and that is precisely what you are wearing. Moreover, the police officer has been told that the suspect has committed an armed assault recently.

The mitigating factor is that the colors you are wearing are not those described to the officer, and the fact that you are black, while the perpetrator described to the officer was white. How does that visual identification change the relative risk?

Where is this going? It depends on the perception of relative risk for each of the individuals. If you are the police officer, you might approach cautiously and undo the flap on your holster while you are still far away. You might call out to the suspect and ask him to stop, and approach cautiously, ready for trouble. You might do any number of things to ascertain the subject's identity.

If you are the young person and if you have committed no offenses or crimes, you might unhesitatingly stop to be questioned, and that would alleviate the officer's concerns about your identity and behavior, or if you are guilty, you might run or take other actions to avoid the officer.

This is relative risk management. What do we know versus what is happening at the moment, and how does that affect us?

Perception of risk is not always rational. It is often the intangible factors that lead us to make a risk calculation. For example, the other night, I was driving home and passed a police car who was traveling below the speed limit. I accelerated to just a little bit above the speed limit in order to get by the police car within the passing zone and was back in the correct lane and well within the dividing line. But the police car turned on its lights and siren and pulled me over.

I did not have any illegal substances in the vehicle, and I was within the law, but the thought that flashed through my mind was myriad and analytical. Operating without any information until the officer approached and told me what the problem was. As it turned out, I had my headlights set too high, and he was commenting that the headlights needed adjustment or to be kept on low beams when following someone.

Risks are acceptable or unacceptable depending upon one's tolerance for fear of what cannot be controlled. But risks are everywhere. Fear of flying is just one of many types of risks we encounter every day. Some of these risks are pointed out in Table 2.1. The purpose is to help put some of the risks in perspective. The data apply to the United States.

The risks we must assess to keep a building, a company, or an industry safe are both internal and external. What we want to do is understand the nature of risks and the structure we use for assessment and look at the different ways we can begin to assign and evaluate risks.

All risks are relative. It is not the risk itself but the perception and presentation of the risk that drive our actions. These actions are influenced by internal and external factors. News, rumor, personal experiences, etc. all influence our perceptions. Flying is perceived as being more dangerous than driving. Other risks are relative as well. Table 2.2 reflects the perception of risk by the US Public and the USEPA.

The table ranked risks in 2000, but since that time, we have had global cooling, the Fukushima Daiichi nuclear disaster and radiation contamination of foods from that disaster, several tsunamis, earthquakes in California and the predictions of "The Big One" (earthquake), the influence of BPA and other chemicals in foods from the containers, genetically modified organisms, and global warming, to name a few. Whether or not these predicted disasters are real or imaginary, these events change the perceptions of risks because the media help shape our perceptions of risk either by repetition, exaggeration, or both. According to *The Economist* magazine, you

TABLE 2.1   Common Daily Risks[a]

Activity	Type of risk	Probability
Driving a car	Death by accident	1/5,000 (0.02%) per year
Rock-climbing	Death by fall	1:25,000 per hour
Motorcycle riding	Death by accident	1:55,000 per hour
Flying on an airline	Death—all sources	1:1,200,000 per hour
Living	Death from heart disease (annual)	1:340
Living	Death by murder (annual)	1:11,000
Living	Death from cancer (annual)	1:500

[a]From Robert R. Johnson. *Background Information: A Scientific View of Risk.* User-Centered Technology (Suny Series, Studies in Scientific & Technical Communication) Paperback—October 29, 1998. Oxford (OH): Center for Chemistry Education, Miami University. www.terrificscience.org.

TABLE 2.2   Relative ranking of perceived risks[a]

Relative risks as perceived by the US Public and by the US Environmental Protection Agency

Risks ranked from highest to lowest by the public	Risks (unranked list) by the USEPA
Hazardous waste sites	Global warming
Industrial water pollution	Urban smog
On job exposure to chemicals	Ozone depletion (caused by CFCs)
Oil spills	Toxic air pollutants
Ozone depletion	Alteration of critical habitats
Nuclear power accidents	Biodiversity loss
Radioactive wastes	Indoor air pollution
Air pollution from industrial sources	Drinking water contamination
Leaking underground storage tanks	Industrial chemical exposure

[a]From Nebel B, Wright R. *Environmental Science.* Elsevier; 2000. p. 403.

have three times the risk from suicide (1:8447) than you do from an assault by a firearm (1:24,974).[1]

We are frightened of new things and especially new risks because of our unfamiliarity with them or their causes. However, risks tend to age poorly, and our perception of the risks is affected by our longevity with them. How many times have you experienced, "It will never happen here because I've been at this plant for X years, and it hasn't happened yet!" While true, the statement provides false assurances and should not influence the risk assessor to derate the risk nor its potential consequences.

SECURITY RISK ASSESSMENT STRUCTURE

Risk assessment documents can take many forms depending upon the type of information one wants to consider. Almost all the assessments consist of tables that summarize the risk in terms of facilities and cost.

The regulatory community often seeks to reduce public risk through increased regulation. That is slightly different than our focus, but it may be instructive even if only from the standpoint of the "law of unintended consequences." Two quick examples will suffice:

> According to the National Highway Transportation Board Report Number DOTHS809835, the mandated new technologies for passenger cars and light trucks issued between 1968 and 2002 cost an estimated $750,782 dollars per life saved in 2002 by implementing these newer technologies.

> The proposed rules on installation of new backup television cameras in cars and light trucks could save 292 lives, on average, per year, and the cost of each of those lives saved would be approximately $18.5 million dollars.[2]

VALUE AT RISK

Industry generally uses a more direct type of risk benefit analysis. It is often simply called risk analysis or risk assessment, and the scope, the focus, and the costs are generally better than most government figures because industrial practice forces constrain the estimates to be relevant, reasonable, and focused on the facilities and the outcomes of specific events on those facilities. This is much more constrained than the value at risk (VaR), which may include market share and financial risks.

Table 2.3 and Figures 2.1–2.3 present three types of risk assessment forms in current use. Each has their advantage and disadvantages. There is no one "right" form for data presentation.

Priorities are generally assigned by vulnerability (column) and frequency (top row). If a structure or an event is highly vulnerable and the frequency is high, it will be in the striped zone and should receive priority consideration. Costs are associated with these events and are generated separately for presentation.

SANDIA LABORATORY'S RISK ASSESSMENT ANALYSIS

The Sandia National Laboratory suggests that threats be categorized in two very specific ways. The first is commitment attributes, which measure the attacker's intent or willingness. The second is the ability or resource attribute, and that is a measure of the attacker's ability to execute the intent.

The intent attribute is further classified into *intensity, stealth, and time.* The first two are measured on a high/medium/low scale, and the time is measured on the immediacy of the planned activity, which can range between days and years.

The ability or resource attribute also has three components, *personnel, knowledge, and access.* The personnel category has several ranges depending upon the number of people who can be applied to the task of planning or executing an attack: *hundreds, tens of tens, ten, or ones.* The knowledge category has two major groups, *cyber knowledge and kinetic knowledge.* Each of these is further refined into high, medium, and low categories.

The net result of this type of planning is a threat matrix table shown in Table 2.3.

The threats are ranked in order of their significance when the resources are accounted for in each of the categories. The purpose is to help delineate the threat categories and separate the rumored threats or theoretical threats from the actual threats.

There are modifiers for the threats that are known as force multipliers. These are funding, assets, and technology. Funding is a critical element and traditionally

TABLE 2.3   SANDIA National Laboratory risk assessment table[a]

				Resources			
	Commitment				Knowledge		
Threat level	Intensity	Stealth	Time	Technical personnel	Cyber	Kinetic	Access
1	H	H	Years to decades	Hundreds	H	H	H
2	H	H	Years to decades	Tens of Tens	M	H	M
3	H	H	Months to years	Tens of Tens	H	M	M
4	M	H	Weeks to months	Tens	H	M	M
5	H	M	Weeks to months	Tens	M	M	M
6	M	M	Weeks to months	Ones	M	M	L
7	M	M	Months to years	Tens	L	L	L
8	L	L	Days to weeks	Ones	L	L	L

The shades indicate relative importance the deeper shading indicates higher priority.
[a]From Duggan DP, Thomas SR, Veitch CKK, Woodard L. Categorizing threat: building and using a generic threat matrix. SANDIA Report, SAND2007-5791. Livermore: Sandia National Laboratories; 2007.

Item description	Vulnerability rating	Criticality rating				
		Very high	High	Moderate	Low	
	Very high	▨	▨	⣿	⣿	
	High	▨	▨	⣿	⣿	
	Moderate	⣿	⣿	⣿		
	Low	⣿	⣿			

Figure 2.1 Classical risk assessment form. The risk analysis matrix is usually in color. Red indicates high risk, yellow indicates moderate risk, and green indicates lower levels of risk, but we have chosen to use stripes, dots, and white spaces to highlight the risk levels, respectively.

reflects the idea of capability, but if the attacker uses the funding to gain outside resources or make payments for information, those acts can make his activities more visible, decreasing his stealth and surprise element.

Assets are the ability to gain other forces or multiply his forces and capabilities. Gaining other assets can also involve the introduction of outside forces. "Two people can keep a secret if one of them is dead" was first voiced by Ben Franklin, and it exemplifies the idea that the more people are involved in a plot, the greater the chance that it will be discovered. Gaining assets may also serve to reduce stealth and be self-defeating.

Technology is rapidly changing. Is the company at or ahead of the technology curve, or does the adversary have the ability to penetrate and defeat the company's security measures? This is especially true if the company is using unsecured and unencrypted wireless communications for controls and measurements inside the plant or for outside communications.

ANNUALIZED COST ANALYSIS OF RISK

A third type of risk assessment considers cost and frequency in terms of annualized costs of the events. While the table form is the same, the data are in terms of powers of 10, and for the purpose of ease of interpretation, a year is assumed to be 333 days long or (3 years equals 1000 days). Similarly, the costs are in terms of powers of 10, and the costs are generally expressed in terms of millions of dollars of damage. This type of risk assessment requires more work because it estimates the frequency and the damage from an event.

> Annual loss expectancy/estimated replacement cost
> Cost expressed as $X.XX \times 10^N$, Rating $= N$
> Frequency of occurrence of undesirable event
> (3 years approximates 1000 days)

1/300 years	$f=1$	Type of event
1/30 years	$f=2$	Type of event
1/3 years	$f=3$	Type of event
1/100 days	$f=4$	etc.
1/10 days	$f=5$	
1/day	$f=6$	
10/day	$f=7$	etc.

Calculated annual lost expectancy $= 10^{(f+N-3)}/3$

In this model, one way of calculating annual loss expectancy may require updating construction cost estimates. If the plant is old, a new engineering-based cost estimate may be required. If the replacement cost estimate is in the last 10 years, *Engineering News-Record*, RS Means, or other construction cost indices may be used to update the cost to the present.

Yet, another method of presenting annual cost data would depend upon the historical database and the confidence that one has in the predictions.

If you believe that it will be 30 years until the next major hurricane that would damage or destroy the plant, then use a 30-year capital recovery cost factor. If the plant will be destroyed in 30 years, and the annual interest rate plus cost of inflation is 8%, then the capital recovery factor tables use the following formula:

$$\text{Estimated annual cost} = \text{present estimated cost} \times \left[\frac{i(1+i)^n}{(1+i)^n - 1} \right]$$

If the replacement cost of the plant in today's dollars is $10,000,000, the time frame is 30 years, and the interest rate is 8%, then the estimated annual cost is $10,000,000 $\times$ 0.0888 = $888,274.

The problem with this is that the cost projections, even for partial damage, require an additional computation and are dependent upon the confidence one has on one's ability to predict the likelihood of adverse events with any degree of accuracy. If the database is reliable, the method of predicting the method will reflect costs more accurately. This is true no matter what method you use to estimate risk. The ability to forecast future replacement costs is often more art than science, and is subject to interpretation and analysis (Figure 2.2).

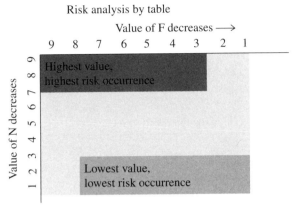

Figure 2.2 Cost-based risk assessment for annual loss expectancy.

SCENARIO-DRIVEN COST RISK ANALYSIS[3]

Scenario-driven cost risk analysis is a way of looking at risk costs using a simple technique to develop the ranges of costs associated with risk. The procedure calls for several scenarios to be developed and costed. The minimum number of scenarios to be analyzed is two, but more are better. Of these scenarios, select the one considered most critical to guard against, and refer to it as the prime scenario. The procedure for scenario-driven cost risk analysis follows in the steps below:

- *First:* Start with the baseline or base replacement costs for the system or unit you are considering. Use current cost estimates for the value of the facility, adjusted to current dates, as if you were going to replace the facility as brand new. Do not allow any adjustment for risk or replacements. Define this as the base cost, or C_b.
- *Second:* Define the prime scenario as the cost elements adjusted for the risk against which you want to guard. Be sure to include all costs and replacement and cleanup costs, including loss of product and other associated costs. Define this as C_{ps}. Note that the damage and cleanup costs as well as replacement costs for the unit damaged should be included. Make sure that you have not incorporated other elements, especially support elements into the replacement and remedial costs.
- *Third:* Subtract C_{ps} from C_b. This is a measure of the amount of reserve money needed to guard against the prime scenario: $C_g = C_{ps} - C_b$.
- *Fourth:* Assume that a measure of probability D will fall between C_{ps} and C_b. This establishes the upper and lower bounds of your cost estimate.
- *Fifth:* Assume that the statistical distribution for the cost lies within the interval of D and that the cost for D is normally distributed between C_{ps} and C_b. D is a probability percentage expressed as a number less than 1, that is, $0.0 < D < 1.00$.
- *Sixth:* Assume that the minimum cost for the system will be $C_{min} = C_b - C_g \times (1 - D)/2$ and the maximum cost for the system will be $C_{max} = C_{ps} + C_g (1 - D)/2$. This establishes the range of costs for probable upper and lower costs.

This gives you four data points:

- C_{min}, C_b, C_{ps}, and C_{max} and a probability (you selected) that the actual number will be somewhere between C_{min} and C_{max}. The probability that $cost \leq C_{max} = 1/2 + 1/2D$.
- The probability that $cost \leq (C_{max} + C_{min})/2$ (average of your estimate of C_{min} and C_{max}) = 1/2.
- The mean of the $cost = (C_{max} + C_{min})/2$.
- Variance of the $cost = \sigma^2_{(Cost)} = (1/12) \times [(C_{ps} - C_b)/C_b]^2$.
- The probability that the cost will be equal or less than a specific figure is

$$\text{Probability}(cost \leq X) = \frac{X - C_{min}}{C_{max} - C_{min}}$$

where X is a real number in currency, as are the other figures.

Real-world example

If C_b in Table 1.1 is $44.3 million, we can estimate that the total cost of the plant replacement with cleanup from a devastating incident would be $65 million. Assuming further that the probability of an attack might be 80% or $D=0.8$, C_{min} would be $44.3 - (75 - 44.3) \times 0.8/2 = 32.02$ million:

$$\text{The maximum cost } C_{max} = 75 + (75 - 44.3) \times \frac{0.8}{2} = 84.28 \text{ million}$$

The equation plotting the cost for the scenario is

$$\text{Probability for } X - \frac{34.43}{84.87 - 44.3}$$

It is linear over the range of the costs, and the table and graph for our probability plot of costs are Table 2.4 and Figure 2.3:

So under the assumptions above, at an 80% risk level, there is a 70% chance that the total costs for the scenario we described is equal to or less than $65 million.

This represents one way of evaluating the financial risk. The technique can also be applied to subsystems.

MODEL-BASED RISK ANALYSIS

The model-based risk analysis (MBRA) is a way of prioritizing costs for reducing risk. The program was developed by the Naval Postgraduate School Center for Homeland Defense and Security (http://www.chds.us). The MBRA is a risk assessment model that takes a network approach to risk assessment.

The tutorial is excellent and easy to use, and the program is clear. While it is designed by and for the United States, other maps can be input and the model can be

TABLE 2.4   **Probability of occurrence**

Estimated cost X	Probability of occurrence (%)
35	1
40	14
45	26
50	38
55	51
60	63
65	75
70	88
75	100

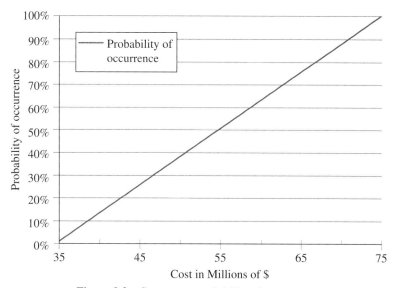

Figure 2.3   Cost versus probability of occurrence.

run without mapping. In the program, one assigns nodes and links, the likelihood of an attack, and the total amount of money that can be sent. The program can allocate and prioritize the amounts of money to be spent on each improvement. Sample inputs to outputs from the MBRA program are shown below.

MBRA example case

The MBRA test case was created to simulate the material flow in an ammonia plant. Artificial numbers were created using hypothetical inputs. The example followed the MBRA tutorial. The hypothetical example was based on the Ammonia Plant example in the previous chapter. The flow of materials in the plant is shown in Figure 2.4. Tables 2.5 and 2.6 show the allocation of resources for an attack scenario, and Figure 2.5 shows the allocation of resources to minimize the damage from an attack and prioritize the expenditures.

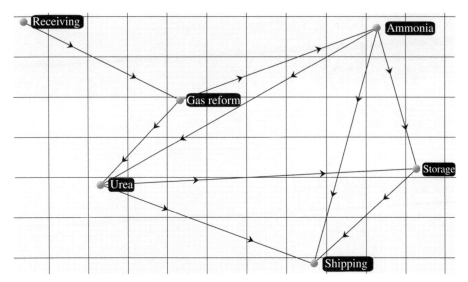

Figure 2.4   Diagram of product flow in an ammonia plant.

In the table above, the data entered are shown in a shaded background (Tables 2.5 and 2.6). The balance of the data is calculated based upon those values. MBRA has allocated the proposed expenses to reduce the risk of the various units of the facility. It also calculates the reduced vulnerability from the expenditure of a portion of the budget for the proposed enhancements of the facility. It is a good guide, but alas not perfect. The problem is that projects have finite boundaries, and a planned upgrade costs what it costs, and cannot be shaded by a risk management program. So if the planned enhancements to a particular area cost more than the computer-allocated expenditures, then the budget will have to be adjusted by cutting things elsewhere.

The other significant feature of MBRA is the ability to prioritize the important links and nodes for reduction of risk. This is shown in Figure 2.5. The figure illustrates the significant links and relative importance of each of the links and nodes. Of course, the data are the limiting factor, and this also illustrates the limits of the program. In the ammonia plant example, the receiving is only a pipeline. The gas reformer and the ammonia conversion and urea conversion units are the heart of the plant and need to be protected. Urea cannot be made without ammonia, and the shipping and storage departments are relatively dispersed. The MBRA program can also analyze risk using the fault tree method.

RISK MANAGEMENT BY FAULT TREE METHODS AND RISK-INFORMED DECISION MANAGEMENT

Fault tree analysis

Fault tree is a method of diagramming and assigning probabilities for risk of complex events by breaking them down into logical steps. The diagram, when complete, is very much treelike in that it has a single event (attack), and the steps that lead up to the attack

TABLE 2.5 Part 1 of two-part data table for MBRA analysis

Threat	Vulnerability	Consequence ($)	Name	Prevention cost ($)	Response cost ($)	Risk initial ($)	Risk reduced ($)	Flow consequence ($)
100	100	30	Receiving	15	75	10	0.48	10
90	100	150	Gas reformer	200	40	9	0.48	10
90	100	80	Urea	45	20	0	0.05	0
100	100	200	Ammonia	500	15	0	0.05	0
100	100	60	Storage	300	2	3.33	0.17	3.33
100	100	85	Shipping	85	300	10	0.48	10
100	100	10	Incoming gas	15	15	10	0.32	10
100	100	50	Gas/ammonia	20	50	0	0.05	0
100	100	0	Ammonia/urea	0	0	0	0.05	0
100	100	10	Reformer/urea	15	2	0	0.05	0
100	100	40	Urea/storage	25	60	0	0.05	0
100	100	12	NH$_3$/storage	10	6	3.33	0.13	3.33
100	100	300	Storage/out	55	250	6.99	0.33	6.99
100	100	5	NH$_3$/out	15	25	0	0.05	0
100	100	20	Urea/out	10	40	0	0.05	0

TABLE 2.6   Part 2 of two-part data table for MBRA analysis

Name	Prevention allocation ($)	Response allocation ($)	Attack allocation ($)	Vulnerability reduced	Consequence reduced ($)	Calculated threat
Receiving	15	0	15	5	10	95
Gas reformer	200	0	200	5	10	95
Urea	0	0	0	100	0.05	90
Ammonia	0	0	0	100	0.05	100
Storage	0	2	0	100	0.17	100
Shipping	85	0	85	5	10	95
Incoming gas	15	1.98	15	5	6.73	95
Gas/ammonia	0	0	0	100	0.05	100
Ammonia/ urea	0	0	0	100	0.05	100
Reformer/urea	0	0	0	100	0.05	100
Urea/storage	0	0	0	100	0.05	100
NH_3/storage	10	0.43	10	5	2.69	95
Storage/out	55	0	55	5	6.99	95
NH_3/out	0	0	0	100	0.05	100
Urea/out	0	0	0	100	0.05	100

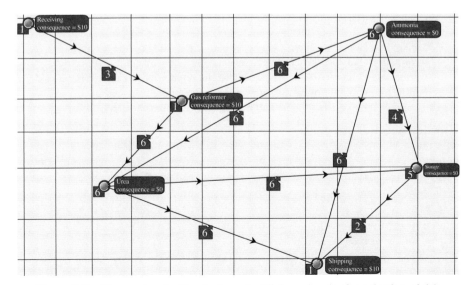

Figure 2.5   Diagram to prioritize the important links and nodes for reduction of risk.

are broken down in fine detail. Those of you who are familiar with project evaluation and review techniques (PERT) or critical path management (CPM) construction management techniques will be extremely comfortable with fault tree or event tree or root cause analyses. Event tree analysis (ETA) and root cause analysis techniques are focused on the past. The fault tree analysis (FTA) is forward-looking, trying to anticipate how things might occur. ETA, FTA, and RCA focus is on the past, seeking to understand; FTA is forward-looking, trying to understand what could occur or go wrong.

All of these techniques focus on a stepwise analysis of the logical progression of events. All of the techniques are binary, in that while there may be multiple events feeding a particular condition, that fault will either occur or not occur and pass through to the next level or to a conclusion.

When ETA is performed at the same time as the FTA with a central event, the result is generally referred to as bow-tie analysis. In most bow-tie analyses, the event is the central item that is displayed in graphical form, as will be shown later, and the faults are plotted to the left and the events are plotted to the right. The advantage of bow-tie analysis is that it displays faults and barriers in one graph. That will be covered later.

RIDM

RIDM is an acronym for risk-informed decision management, and this and continuous risk management (CRM) were used by NASA for risk management. The overall NASA formula for risk management is

$$RM = RIDM + CRM$$

Quoting from a NASA document[4],

> RIDM is a fundamentally deliberative process that uses a diverse set of performance measures, along with other considerations, to inform decision making. The RIDM process acknowledges the role that human judgment plays in decisions, and that technical information cannot be the sole basis for decision making. This is not only because of inevitable gaps in the technical information, but also because decision making is an inherently subjective, values-based enterprise. In the face of complex decision making involving multiple competing objectives, the cumulative wisdom provided by experienced personnel is essential for integrating technical and nontechnical factors to produce sound decisions.

Risk management by NASA's standards is a continuous process because of the levels of uncertainty with which they deal. We have chosen NASA's risk management process for further explanation and evaluation because many of the uncertainties in the security process and in NASA's work product are similar in that they deal with many unknowns. NASA may have one of the best and most effective risk management systems, as it focuses on continuous improvements, and once mastered, it will lead to an understanding of many different types of risk management systems.

NASA defines risk as an "operational set of triplets," scenarios, likelihoods, and consequences. NASA further defines the RIDM process as a set of continuous interactions between the stakeholders, risk analysts, subject matter experts, technical authorities, and decision maker. The RIDM process is conducted in the same manner as the conventional risk assessment processes. Figures 2.6, 2.7, and 2.8 illustrate the overall process.

The RIDM process is very much like the ISO 9000, 14000, and 18000 processes and related systems, where the process evaluation and review is continuous until the process is either complete or that a consensus of optimization has been achieved.

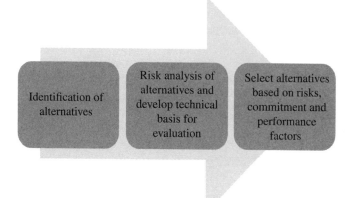

Figure 2.6 NASA's risk-informed decision management process. From *Probabilistic Risk Assessment Procedures Guide for NASA Managers and Practitioners.* 2nd ed. NASA/SP-2011-3421, NASA; 2011.

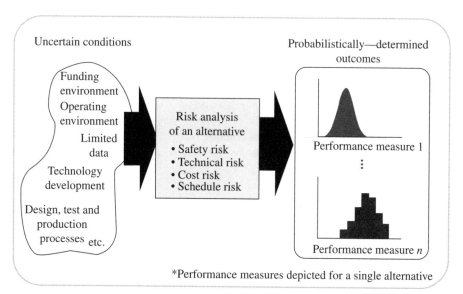

Figure 2.7 Factors that go into a risk-informed decision management process. From Stamatelatos M, Dezfuli H. *Probabilistic Risk Assessment Procedures and Guidance for NASA Managers and Practitioners.* NASA/SP-2011-3421, NASA; 2011. http://www.hq.nasa.gov/office/codeq/doctree/SP20113421.pdf.

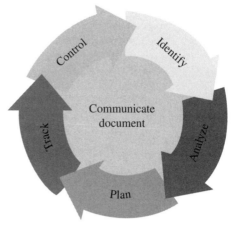

Figure 2.8 Steps in the RIDM process. From *Probabilistic Risk Assessment Procedures Guide for NASA Managers and Practitioners.* 2nd ed. NASA/SP-2011-3421, NASA; 2011.

That consensus would find that the process is complete and that the optimum balance between risk and other factors has been achieved. The RIDM process was specifically designed for production or mission-related activities by NASA, such as shuttle launch decisions and manufacturing special parts and systems, but the processes are the same for a security system.[5]

The International Atomic Energy Agency has adopted the RIDM process and adapted it to nuclear works. The similarities are obvious and are shown in Figure 2.9.

RIDM process steps   The steps for the RIDM process are as follows.

Step 1: Identify   Capture stakeholder's concerns regarding the performance requirements and or the critical elements that must be protected. (Note: the following section on CARVER + Shock may have some very good suggestions about ranking and setting priorities.) Each element must have an associated risk, and the probability of the success or failure is dictated by one or more scenarios. It is the consequences of the risk scenario that determines the outcome.

As an example, if we say that the gas reformer maximum scenario is the complete destruction of the reformer and the resulting shut down of the plant, then anything less is a partial success, and we can attach numbers to that in terms of percentages of shutdown and/or production lost of the other operations as well.

Other scenarios and multiple event scenarios are equally likely. There are lots of things that can take place, and some of them are routine, internal, or external and are often considered during the design stage. Examples might include power interruptions, an electrical voltage spike, simple corrosion, advanced corrosion due to electrolysis, and wear and tear on the system. These are technically "attacks," but their nature and response are more in the realm of design deficiencies and maintenance and repair activities.

Sabotage, in large or small scale, is just one type of attack. Technically, there is little difference between sabotage and "mischief" except in the damage done. Mischief may be as simple as disabling equipment so that an employee gets additional

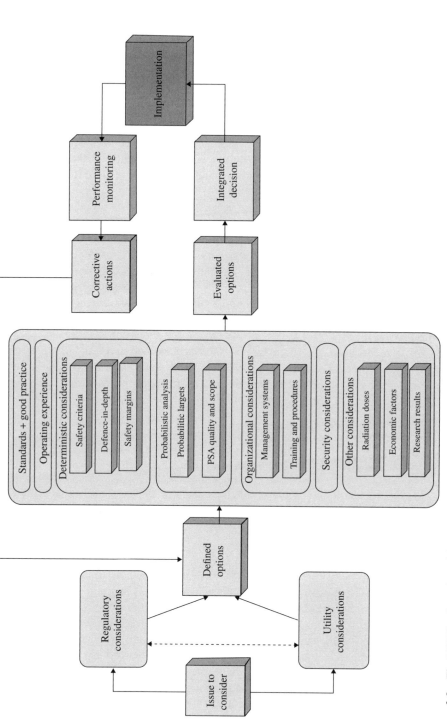

Figure 2.9 The IAEA's adaptation of the RIDM process. From IAEA, A framework for an integrated risk informed decision making process. INSAG-25.

overtime pay to plugging a vent just before the plant union goes out on strike to make it difficult to clean out a chamber. Sabotage can create actual and willful harm to major elements of the operating portion of the plant. The difficulty is that mischief is difficult to detect and is more of an inconvenience.

Other types of scenarios may include physical or other electronic interference with remote locations or communications. A remote pumping station is a vulnerable area easy to attack. The scenario might be just someone breaking in and stealing components or simply cutting the power or massive vandalism.

Any teenager with some talent and a modem and a wire wrapped around an orange juice container can make a transmitter, which could send signals to the unsecured transmitters and receivers, which open and close valves or control chemical or thermal reactions.

While some of these scenarios are very simple, the threats posed can be quite complex. The point is that they need to be considered in the planning and posed as potential occurrences.

In the overall preparation of scenarios, it is important to consider the effects of any environmental or other effects from unplanned releases of chemicals within the plant. If incidents within the plant result in the release of volatile chemicals the community notifications and environmental regulatory notifications may be required. There are several good programs that can assist in the planning for various types of environmental releases.[6]

NASA, because it is focused on specific missions, recommends that the scenario development be decomposed into a set of steps of consulting with the stakeholders to gain their concurrence that the scenario is realistic. Quoting from their RIDM document:

> Stakeholder expectations result when they i) specify what is desired as a final outcome or as a thing to be produced and ii) establish bounds on the achievements of goals (these bounds may for example include *costs, time to delivery, performance objectives, organizational needs*). In other words, the stakeholder, expectations that are the outputs of this step consist of i) *top-level objectives and ii) imposed constraints*. Top-level objectives state what the stakeholders want to achieve from the activity: these are frequently qualitative and multifaceted, reflecting competing sub-objectives (e.g., increase reliability vs. decrease cost). Imposed constraints represent the top-level success criteria outside of which the top-level objectives are not achieved.

So, develop your scenarios as completely as you can, and involve various departments on the scope of the scenario. Get consensus where possible, especially if it involves outside notifications and support from others.

Step 2: Analyze   The analysis step requires the estimation of the magnitude and consequences of individual risk elements and working through scenarios to completion. These are the consequences of the scenarios in step 1 and should include recovery steps and related costs until the plant is fully restored and operating.

One of the chief problems in the analysis is related to the timing of the attack or the probability of success. If there are *n people in the scenario committee, anticipate that the number of attack and timing scenarios to be evaluated will be between 120 and*

250% of the number of people you have on your committee. These opinions are just that, and the essential difficulty in production of a coherent report is that you do not know what you do not know, and you may not ever know that you do not know it until after the attack has occurred. Without good intelligence from the community, you will never know whose scenario and timeline about an attack is more accurate until after it occurs.

The second area of possible disagreement is the assessment of the damage. One persons' minor damage is another persons' total destruction. Here is where you must involve the process engineering group with designations of the key elements of the plant, which are vulnerable to explosions, fire, weather, control failures, etc.

The third area will be the cost of reconstruction. If the plant has been built in the past 5 years or so, it might be possible to utilize a Construction Cost Index for estimation of the cost of replacement. Otherwise, you could be looking at a much more comprehensive engineering study to estimate the costs. Vendors can and will provide quotations for their good clients. It is also possible to get some indication of the cost of replacement by comparison of published data on comparably sized new equipment. Chemical engineering handbooks have some guidance on scaling up capital costs for process equipment.

The preparation of a detailed cost estimate for a facility can be several hundred to several thousand hours of estimator's time just to get within 25% of the actual cost, depending upon the level of detail required, the complexity of the plant or process, and the accuracy of the estimate required.

We never have sufficient data to predict when an attack might occur, and a continuous vigilance is required, with the possibility of extreme rapid response when and if an event occurs. There are scenarios and then there are extreme scenarios, and the response must be proportionate to the type and kind of attack.

The analysis of the effects of an attack depends upon the scenario. Some attacks will occur quickly, and others will provide warning. Adverse weather, flood, storms, and weather-related events provide warning of their occurrence and severity. Other attacks will not provide warnings and their severity can be much greater. A fire in the pump house may have minor consequences, or it may lead to a plant-wide conflagration. It is sometimes difficult to ascertain the difference between plant safety issues and plant security issues. However, for the purposes of our discussion and planning, an attack will be an external event, caused by outside forces or outside persons.

Confining ourselves to events attributed to outside forces still leaves us with a problem related to the estimation of the timing and the duration. A blizzard or bad ice storm may cause major disruption, or it may be a minor event that will disrupt the power. A physical assault by a terrorist organization or a cyber attack designed to disrupt processes or steal information or divert shipments or conceal other misdeeds can be difficult to analyze because one does not know the scale of the event. In the case of the theft or diversion, the detection is often after the fact.

Sabotage as opposed to mischief is also an attack. Someone deliberately leaving a door or gate unlocked to allow an intruder into the facility could be classified as an attack or it could be simple theft or sabotage. It is hard to tell beforehand what can happen unless one gets down to the area-specific planning of intrusion attacks. Someone in the tank farm area may be harmful or not, whereas an unauthorized someone in the computer center probably will be a potentially serious event because of the harm that can be done.

The analysis process will have multiple outcomes based on each scenario. The timing and nature of an attack is a best guess. Where there is information available from police, military, or reliable intelligence sources, the timing of an attack may be known or estimated. Otherwise, the system must be ready for eternal vigilance.

The overall process of analysis includes several important elements:

- What can go wrong—or how can an attack occur?
- How big would the attack be?
- What are the security systems to prevent, deter, or defeat an attack?
- What is the margin or reserve to prevent an incident or follow-up attack?[7]

Obviously, the attack scenarios must be limited to those that have a realistic chance of occurring. Otherwise, a plant on the equator would be evaluating the possibility of being hit with a blizzard.

In planning the scenarios, you want to include the ultimate scenario even if it does seem remote. Unless you are confident that the likelihood of an occurrence of the ultimate destruction scenario is less than about 1 in 10 million or less, you would probably want to include the ultimate disaster scenario in your planning. That should, of course, include the ultimate destruction of the facility by explosion or fire or both, and community evacuations and cleanups.

Step 3: Planning   This is the step where the response to the attack scenario is plotted out. Logistics should include the manpower required for the type of response and should be specific to the scenario. Options tend to be the enemy of detailed planning but are necessary. It is an exercise in visualization in that one has to (i) put themselves in the place of the attacker, and (ii) if the attacker is detected, estimate how much force is required, or (iii) if the attacker is not detected, estimate what could or would be damaged by the attacker.

One of the easiest considerations in this type of exercise is to estimate how long the attacker can operate before the attack is detected. If, for example, an attacker manages to penetrate the outer perimeter of the plant, how far can he get before detection occurs and the plant is alerted? Think of the attacker's presence as a wave spreading out from a pebble tossed into a pond. Gates, locked doors, fences, etc. are not necessarily deterrents, but they will slow an attacker down, making the area of probable access and search radius smaller.

Other questions in the response planning scenario include:

- What resources are required?
- How will outside resources be used, including fire, police, and emergency services?
- How fast can these resources, both internal and external, be tapped, and what should their response be?
- Are the equipment or personnel prepositioned so that they can respond rapidly to the event?

One of the other critical elements in the response planning scenario should include the availability of emergency services such as a hospital and its response time. A part of that series of questions should include analysis of the directions and transit times, and the ability of the hospital to handle or decontaminate injured personnel without damage to their emergency room or other facilities?

Sample planning exercise   Another way of looking at the planning exercise is to consider the types of elements outlined below:

Start here

Deterministic element
Defense in depth
Barriers and levels

Security margins
Flexibility of defense force
Experience and history
Communications systems
Backup communications

Exercises and drills
Outside resources

Probabilistic elements
Provide analysis of likelihoods of various penetration and attack scenarios
Can the facility handle multiple events for both attack and disasters if they happen concurrently?

What are the notification and compliance requirements?

Continue here

Deterministic approach
Answer the question:
Are the existing security systems meeting their required purpose and function?

If not, what are the consequences?
What are the constraints on the systems?
What can be done to improve the systems?
How much will the improvements cost?

Probabilistic approaches
What can happen with a security breach or an attack?
How serious could it be?
How likely is it to occur?
How reliable are the outside resources?
What are the capabilities of the outside resources to handle a catastrophe?

Compare both sets of answers to target goals to see if the security is adequate for the facility

If yes, you are done; now, monitor the system

If no, start back at the beginning and reanalyze the system

The company's tolerance for risk should be addressed, as it may be a determining factor in the overall security planning. In most cases, the risk statistics will be low, and the risks are probabilistic rather than actual.[8] Risk levels, based upon experience, are generally of little use, because the history will likely suggest that the "attack" did not happen. "Did not happen during previous history" does not suggest that the "attack" cannot or will not happen. The probability maybe very low, and one will be trying to decide between the likelihood of an event with a time based probability of 0.001 and 0.0001% or less. One has to determine the tolerance for risk, even for small catastrophic events, as it will affect the resources available.

The risk management decisions should be elevated to the sponsoring organization at the next higher level of hierarchy when the risk cannot be managed by the present organizational unit. If the risk is plant-wide and may have major consequences on the corporation, the decision about levels of risk should be sent to the appropriate corporate management level for their sign-off or commitment. Risk can be no longer considered when the associated risk drivers are no longer considered potentially significant.

Step 4: Track All risk decisions must develop a paper trail so that they can be examined and evaluated after an exercise, after an event or even after a failure. The purpose is not to see who failed, but to find out how the system worked and how well it worked in preventing the event. The meetings and documents should be captured and recorded so that the thought process of the risk managers can be evaluated and captured for later consideration.

Unfortunately, this can result in huge paper and electronic files. The most effective "paper trail" will be a time-stamped set of electronic communications files followed up with written reports every time there is an incident, no matter how small. This documentation should include copies of video files and radio communications both internally and external to the plant. Especially where there is an injury or an arrest, it will be vital to the defense of the plant and employee to insure that actions are defensible and the appropriate level of force has been employed.

Step 5: Control Periodic review and editing of the risk management plans are required. Optimally, this will result in a book of procedures that is used in training the security force and key managers. Optimally, there should be a set of training requirements that relate to the procedures. This training document should specify the levels of education, positions, qualifications, and examinations for each level of security personnel. The procedure manual does not need to be highly detailed, but it should be specific enough so that the security force personnel know what actions to take and what appropriate levels of force, response, and reserve are required. The procedure manual should also specify the qualifications, responsibilities, duties, and authority for each level of the security force.

Much of the language in the procedure manual will be repetitive, and it may be helpful to have some prewritten descriptions of standard actions available to plug into the documents. *CAUTION: This type of standard language can lead to unimaginative and preprogrammed actions where the plan is not taken seriously or is put on the shelf as an unused reference work, only to be used as a justification of actions taken whether they were appropriate or not.*

If an armed guard force is used, the plan must specify the levels of force and specify the training requirements and the qualifications for the security officers before they are allowed to be armed. If the guards are not armed, the plan should specify the procedures for contact of the local police and/or fire department and the levels of supervisory control who can make that contact.

Ideally, the review process should be a continuous or semicontinuous basis. The security manager and/or the security team should review the existing plans on a regular basis. Depending on the size of the plans and the complexity of the facility, that may require almost continuous review and updating. But the plan needs to be reviewed at least yearly, or more frequently wherever possible.

Every incident needs to be reviewed to insure that the response was appropriate to the incident and that the plan is revised to insure appropriate force/response levels are employed. If a portion of the response or risk management plan becomes untenable or nonviable, it should be stricken from the master planning book.

The universal limitation of all the planning techniques we have discussed so far is their inability to tell us when an incident could occur. **What we do not know, we do not know**. The solution is planning and eternal vigilance against the unplanned, accidents, and incidents.

RIDM procedures   The procedures for preparing decisions under the RIDM process start with the preparation of boundary statements and objectives.

- The objective of the plant security is to protect the facility and prevent damage, theft, sabotage, external aggression, and loss from other internal or external sources.

- The security operations will be responsible for all of its own computer and related communications activities separate from the computer and communications resources employed by the information technology (IT) department.

- The IT department shall support the security department where and when requested.

- The security department is responsible for control of external events and internal threats including the prevention of theft, diversion of assets, and physical intrusion.

- The security department has the responsibility to inspect and control movements into and from the plant, including all shipments.

- The security department will address all potential events that can create a loss event greater than $_____ (a specified sum).

- Those types of statements establish the duties and the responsibilities and boundaries of the security division. The last statement establishes a performance objective or metric against which security can be measured. The next step would be to determine what type of response is required.

- The next step is to establish what type of attack scenarios can create damage and the magnitude of that damage. For example:
  - An intruder in the plant could shut down the reformer.
  - A car bomb.
  - An intruder with a bomb could cripple or disable the power plant.
  - Smuggling a shipment of drugs or explosives into the warehouse.
  - A suicide bomber could take out the front gate.
  - Multiple incidents closely spaced (ripple attack) could provide a diversion and allow multiple entries into the plant.
  - Be sure to include the ultimate scenario for plant-wide destruction in your disaster and attack scenarios.
  - Etc.

Each of these has a damage associated with it, and depending upon the other assumptions associated with it, there may be several possibilities associated with the location, timing, etc. of the event. At this point, a review is needed to select the most likely events based upon the facilities, location, etc. The evaluation of the cost will require input by engineering and other specialists. The scenarios should also be prioritized and ranked with the likelihood of their occurrence and the damage associated with each using the techniques outlined earlier.

There are several ways to arrive at the probability of the occurrence or likelihood of the success of the attack and/or prioritize the target. Another way of predicting the prioritization of the target is CARVER + Shock, which will be discussed next. The FTA can be used to estimate the success or failure of an attack.

CARVER + Shock   CARVER + Shock is a planning tool that aids in developing and selecting priority targets for protection. It was first developed for the food industry, and it is now being expanded into free software that enables other process industries to use it with slight adaptation. The CARVER + Shock system uses a matrix formulation that has already been discussed. The interesting and different element is that it also considers the shock value of an attack and the damage as part of the matrix. The word CARVER is an acronym for *c*riticality, *a*ccessibility, *r*ecuperability, *v*ulnerability, *e*ffect, and *r*ecognizability. The Shock part comes in when we evaluate the psychological effect of the attack or incident on the affected facility or nation. It is rumored that during Desert Storm, US Special Operations Forces used CARVER to identify the critical air defense network system in Iraq. The thorough analysis and breaking the system into critical parts identified communications bunkers, which made the functioning of the Iraq's RADAR system possible. When these bunkers were destroyed by small strikes, the larger air campaign against Iraq was made possible with minimal loss of aircraft.

Keep in mind that if you perform a CARVER + Shock analysis, you will need to make several iterations to resolve the scenarios into their smallest parts, which have critical impact upon the facility. As you work through the scenarios, some activities will be important, and others will not have an impact.

Defining the terms for Carver+Shock   *Criticality* is the measure of the public health or economic impacts of the attack. It should be a scaled evaluation of the cost of the attack in terms of budgets, corporation value, facility value, or any other thing.

Accessibility is the ease of ingress and egress to the facility or the site. A site with a fence will have a lower accessibility than one without. Similarly, natural barriers such as rivers, lakes, and natural terrain would give a site a lower score.

Recuperability or recoverability is a measure of the ease with which the facility can be repaired or recovered from an attack on a specific part of the facility or the entire facility.

Vulnerability is the ease of accomplishing the attack. The challenge to assigning a priority or rank to this item is that it also depends upon the type of attack. If a facility is relatively wide open and an intruder can penetrate the perimeter and walk around the facility unchallenged, then the vulnerability would be very high. Even if a facility had a relatively secure perimeter, it might be vulnerable to an attack by a standoff weapon such as a military-grade or homemade mortar.

Effect is the effect that the attack would have, in terms of direct loss, such as direct loss of production capacity or inventory. It might be important to separate inventory from production in this category for separate parts of the facility.

Recognizability is the ease of identifying the target. Much of that may depend upon the attacker's familiarity with the plant and his ability to distinguish high-value targets from those having little or no value. A good example of a high-value target might be the gas reformer in an ammonia plant versus the bulk gas or liquid storage tanks. Often, the tanks will be also protected by a dike or other containment. The loss of the tank integrity if there is no fire, and if the tank containment is intact, may be of minimal consequence.

Shock is an attempt to evaluate the combined health, economic, and psychological impacts of an attack. While it is primarily for the food industry due to the impacts of an attack on an individual, the facility may be insignificant unless that facility is making something of national importance.

CARVER + Shock definitions need to be scaled to suit the plant size and other impacts. The top-end scale should be whatever is agreed upon by plant management, and it is suggested that the upper end of the scale might be, for example, the loss of the entire plant, a large corporation's projected profits for a year, the net worth of the facility, or other appropriate measures. It is important not to set the upper levels of impact too high, and each category should have some 1s and 10s in it. It is important to recognize that there might be a tendency to set values artificially high to stimulate plant investment or to set them too low to avoid inspection and accountability. It is also important to consider that the various categories of attack may produce some casualties that are unavoidable.

Applying CARVER + Shock

Know yourself This includes the entire infrastructure or facility to be assessed. Many other VA systems attempt to identify only critical systems to evaluate in the hopes of saving time and resources. This shortcut can overlook crucial vulnerabilities as proven time and again by users of this tool. Results show that significant vulnerability lies in areas that most experts never considered critical and may have overlooked.

Know the threat

- *Actual, localize threat for a specific target system.*
- Design basis threat for a higher-level assessment.
- We must understand who the threat is, why they want to attack, how they will attack you, and what is the desired effect.

Know your environment This is information about the physical, political, and legal environment that affects the target system and the threat.

Know what your enemy knows about you This is an additional component to this preassessment. Information is sometimes called red teaming. It is not required to identify the actual vulnerabilities; it is used more to predict probability of attack.

CARVER + Shock produces a bar graph that indicates the levels of vulnerability of the facility, and it has a 108-question section interview that discusses recall and processing safety and is primarily designed for the food industry, but with some imagination, it can be used on other platforms and industries (Tables 2.7, 2.8, 2.9, 2.10, and 2.11).

Fault tree analysis Analysis of a failure by fault tree is yet another way of determining the likelihood and/or the probability of an attack. The fault tree system breaks down the planning and activity process into binary actions that have either "go" or "no-go" elements, and when probabilities are assigned to each of the elements, a picture of the overall probability emerges. One of the hardest things to do in an FTA is to decide what the individual probabilities are for the individual actions that lead into events. Monte Carlo methods are ideally suited to the FTA method because the variables can be stepped through to gain an overall probability of success. In order to estimate the probability of an attack, a large number of variables have to be examined, and scenarios run for realistic probabilities. The principal limitations are in the reliability of the assumptions regarding the possibility of an attack and the likelihood of the success or failure of that attack.

The FTA is the most comprehensive type of analysis and can be used whenever there is a logical relationship between events and consequences. Related to FTA is the event tree analysis (ETA). The ETA is the backward consideration of FTA because you are starting with a failure and evaluating various possible causes. FTA is before; ETA is after.

For each event, a fault tree is created using possible outcomes, and using a series of "and" and "or" gates with modifiers, an event sequence is created. While this book is not a fault tree tutorial, it is necessary to provide some background about uniform notation in order to understand FTA construction.

Each element in the event sequence, and each subelement in the event sequence, has a probability so that the event sequence adds up to 1 at each gate: this is symbolized by Π,

TABLE 2.7 CARVER + Shock criticality table

Criticality criteria	Score or ranking
Loss of over 10,000 lives *or* loss of more than $100 billion (Note: if looking on a company level, loss of >90% of the total economic value for which you are concerned*)	10–9
Loss of life is between 1,000 and 10,000 OR loss of between $10 and $100 billion (Note: if looking on a company level, loss of between 61 and 90% of the total economic value for which you are concerned*)	8–7
Loss of life between 100 and 1000 OR loss of between $1 and $10 billion (Note: if looking on a company level, loss of between 31 and 60% of the total economic value for which you are concerned*)	6–5
Loss of life less than 100 OR loss of between $100 million and $1 billion (Note: if looking on a company level, loss of between 10 and 30% of the total economic value for which you are concerned*)	4–3
No loss of life OR loss of less than $100 million (Note: if looking on a company level, loss of <10% of the total economic value for which you are concerned*)	2–1

TABLE 2.8   CARVER + Shock accessibility criteria

Accessibility criteria	Score or ranking
Easily accessible (e.g., target is outside the building and no perimeter fence). Limited physical or human barriers or observation. Attacker has relatively unlimited access to the target. Attack can be carried out using medium or large volumes of contaminant without undue concern of detection. Multiple sources of information concerning the facility and the target are easily available	10–9
Accessible (e.g., target is inside the building but in an unsecured part of the facility). Human observation and physical barriers limited. Attacker has access to the target for an hour or less. Attack can be carried out with moderate to large volumes of contaminant but requires the use of stealth. Only limited specific information is available on the facility and the target	8–7
Partially accessible (e.g., inside the building but in a relatively unsecured, but busy, part of the facility). Under constant possible human observation. Some physical barriers may be present. Contaminant must be disguised, and time limitations are significant. Only general, nonspecific information is available on the facility and the target	6–5
Hardly accessible (e.g., inside the building in a secured part of the facility). Human observation and physical barriers with an established means of detection. Access generally restricted to operators or authorized persons. Contaminant must be disguised and time limitations are extreme. Limited general information available on the facility and the target	4–3
Not accessible. Physical barriers, alarms, and human observation. Defined means of intervention in place. Attacker can access target for <5 minutes with all equipment carried in pockets. No useful publicly available information concerning the target	2–1

TABLE 2.9   CARVER + Shock recognizability criteria

Recognizability	Score or ranking
The target is clearly recognizable from a distance and requires little or no training to identify it	10–9
The target is clearly recognizable and requires a little bit of training to identify	8–7
The target is difficult to recognize at night or in bad weather or might be confused with other targets or target components and requires some training for recognition	6–5
The target is difficult to recognize at night or in bad weather. It is easily confused with other targets or components and requires extensive training for recognition	4–3
The target cannot be recognized under any conditions, except by experts or insiders	2–1

TABLE 2.10   CARVER + Shock vulnerability criteria and effect criteria[a]

Vulnerability: This is a measure of how easy it would be to introduce weapons, bombs, poisons, or other foreign substances into the plant, near the target, or into the target processes	Score or ranking
Target characteristics allow for easy introduction of sufficient agents to achieve aim	10–9
Target characteristics almost always allow for introduction of sufficient agents to achieve aim	8–7
Target characteristics allow 30–60% probability that sufficient agents can be added to achieve aim	6–5
Target characteristics allow moderate probability (10–30%) that sufficient agents can be added to achieve aim	4–3
Target characteristics allow low probability (≤10%) that sufficient agents can be added to achieve aim	2–1

Effect criteria	Score or ranking
Greater than 50% of system's production or function impacted	9–10
25–50% of system's production or function impacted	7–8
10–25% of system's production or function impacted	6–5
1–10% of system's production or function impacted	4–3
Less than 1% of system's production or function impacted	2–1

[a]Criteria established by Carver + Shock Software.

TABLE 2.11   CARVER + Shock shock value[a]

Shock value is the combined measure of the health, psychological, and collateral national economic impacts of a successful attack on the target system. The psychological impact will be increased if there are a large number of deaths or the target has historical, cultural, religious, or other symbolic significance. National economic damage and casualties of innocents (children and elderly) are also factors	Score or ranking
Target has major historical, cultural, religious, or other symbolic importance. Loss of over 10,000 lives. Major impact on sensitive subpopulations, for example, children or elderly. National economic impact more than $100 billion	10–9
Target has high historical, cultural, religious, or other symbolic importance. Loss of between 1000 and 10,000 lives. Significant impact on sensitive subpopulations, for example, children or elderly. National economic impact between $10 and $100 billion	8–7
Target has moderate historical, cultural, religious, or other symbolic importance. Loss of life between 100 and 1000. Moderate impact on sensitive subpopulations, for example, children or elderly. National economic impact between $1 and $10 billion	6–5
Target has little historical, cultural, religious, or other symbolic importance. Loss of life <100. Small impact on sensitive subpopulations, for example, children or elderly. National economic impact between $100 million and $1 billion	4–3
Target has no historical, cultural, religious, or other symbolic importance. Loss of life <10. No impact on sensitive subpopulations, for example, children or elderly. National economic impact <$100 million	2–1

[a]Criteria established by Carver + Shock Software.

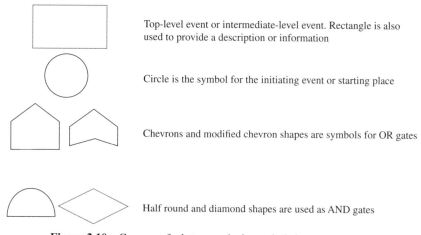

Top-level event or intermediate-level event. Rectangle is also used to provide a description or information

Circle is the symbol for the initiating event or starting place

Chevrons and modified chevron shapes are symbols for OR gates

Half round and diamond shapes are used as AND gates

Figure 2.10   Common fault tree analysis symbols in current usage.

$$\text{where } \Pi = \sum \left(\pi_1 + \pi_2 + \pi_3 + \cdots + \pi_n \right) = 1 \text{ for each gate.}$$

The FTA starts with an event, usually described by a rectangle but defined in a circle. The rectangle is generally used for information or description purposes. OR gates are sometimes chevrons or a rounded chevron shape (shown in the following).

Some of the element shapes are diamonds, sometimes indicating an OR gate and indicating an AND gate. The gates are indicating the likelihood of an event or the probability of the event occurring. Note that for an AND gate, A and B must occur for the activity to pass out of the gate. The presence of A or B will not activate the gate alone.

For an OR gate, A *or* B can cause the fault to propagate through the gate. It is an either or both condition. Consequently, the probability of an event passing through an OR gate is higher than the probability of either of the highest events leading to the gate—because one OR the other or both can take place (Fig. 2.10).

The examples on the next page will help explain the functions.

The mathematical basis for an AND gate is $P_{\text{and gate}} = P_1 \times P_2$ denoted by P_a.

The mathematical basis for an OR gate is $P_{\text{or gate}} = 1 - (1 - P_1) \times (1 - P_2)$ denoted by P_o (Figs. 2.11 and 2.12).

Similarly, if there were three events, on an OR gate, and the probabilities of each threat were P_1, P_2, and P_3, respectively, the probability of a successful attack would be

$$P_{\text{success}} = 1 - \left[(1 - P_1) \times (1 - P_2) \times (1 - P_3) \right],$$

and the probability of failure of all three attacks failing is

$$P_{\text{failure}} = (1 - P_1) \times (1 - P_2) \times (1 - P_3), \text{ etc.}$$

Fault tree analysis

For the ATTACK we have the following:

And Gate:

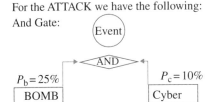

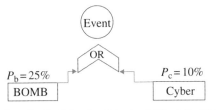

$P_a = (P_b \times P_c) = 0.25 \times 0.10 = 0.025$

If we analyze the probabilities of the system resisting the attack we get the following:

$P_{\sim b} = 0.75; P_{\sim c} = 0.90:$

$P_o = 1 - (1 - P_b) \times (1 - P_c) = 1 - 0.75 \times 0.9 = 32.5\%$

Corresponding analysis for OR is

$P_{\sim b} = 0.75; P_{\sim b+c} = 0.75 \times 0.1 = 0.075$

$P_{\sim c} = 0.10; P_{\sim c+b} = 0.25 \times 0.9 = 0.225$

And $P_{both} = 0.025; \Sigma P = 0.325$

Figure 2.11 Fault free analysis example after Lewis. The example for this was taken from Dr. Ted Lewis work on Network Analysis, Op. Cit.

In some instances, failures can be tracked to equipment or events, and the data on mean time between failures often helps us evaluate the reliability of a particular system. The data on specific items of equipment are available from the American Society for Quality (www.asq.org). For certain classes of equipment, there is a MIL-STD 781-C "Reviews of Standards and Specifications: MTBF Confidence Bounds Based on Fixed Length Test Results": also available from the ASQ.

Event trees can be used to evaluate the cause of failures. An excellent example of the application of a detailed fault tree is found on the BPs analysis for the cause of the Deepwater Horizon failure[9]. The Deepwater Horizon Accident Investigation Fault Tree report has been deleted from the BP website and replaced with a discussion of the implementation of their "internal" report findings and recommendations to improve safety, in an attempt to manipulate and polish their poor corporate performance record on safety.

The Deepwater Horizon event was, for that facility, the ultimate disaster scenario, and after paying out several billions of dollars, we are quite sure that BP would agree. The point is that little consideration was given to the possible occurrence of that type of disaster and ultimate failure, and the consequences arising from the well failure, gas release, explosion, and fire, and the failure of the well-sealing valve.

In hindsight (and hindsight is always perfect), an event tree analysis which considered the consequences of well failure, and the following events, *might have foreseen the events and could have led to installation of safety equipment and better procedures which would have prevented the catastrophe.* The installation of additional controls and procedures which prevented the taking of shortcuts during the drilling operations and greater corporate emphasis and focus on safety and environmental protection *should* have led to a successful development of the well.

Conclusion Risk assessment must be comprehensive and must be a continuous process performed with upper management's knowledge and participation. The

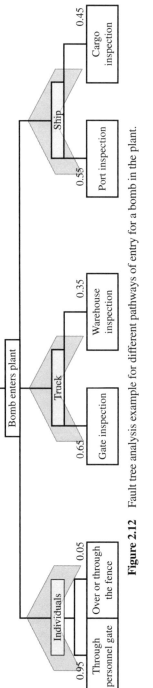

Figure 2.12 Fault tree analysis example for different pathways of entry for a bomb in the plant.

specific method for the assessment criteria is highly dependent upon the type of threat and the method of attack. In most cases, those elements are unknown until after they occur. Therefore, it is necessary to consider a number of scenarios for each important process unit, which addresses worst case, expected case, and minimal case set for the attacks. It is equally important to drill for the worst case and expected case, just as one might expect the plant fire brigade to practice fire drills. Only through planning and rehearsed drills can plant employees successfully handle attacks.

Steps for a good analysis   The steps for a good analysis include upper management participation and a thorough investigation of lots of cases. It is a team effort. Briefly, the steps for analysis are outlined below:

1. List assets. Take an inventory of the assets.

2. Select the type of analysis you are going to use and determine the best and most applicable type of analysis suited to your plant or corporation. Be sure to include an ultimate disaster scenario for each type of plant present at a site.

3. Perform network analysis. Determine hubs and links to find out if there are critical elements that you may have missed.

4. Build a model using fault tree, CARVER, RIDM, MBRA, or other software package.

5. Analyze the model and prepare a fault tree or an event tree to verify the paths.

6. From the event tree, you can calculate the optimum events and impacts to guide you in allocation of resources to reduce the hazards.

7. Circulate the information and get consensus on the inputs, formalize the plan, make upgrades, and use the plan and the scenarios to train the plant personnel.

8. Review the exercises, and then revise the plan as may be required.

NOTES

1 Source: Danger of death. Economist 2013 Feb 14.
2 Source: National Highway Traffic Administration data. See http://www.bloomberg.com/ news/2011-11-17/u-s-rear-view-camera-rule-may-cost-18-million-per-life-saved.html. Accessed 2014 Oct 13.
3 The basis of this approach is from Paul R. Garvey's presentation on "Cost Risk Analysis without Statistics!!" Paul Garvey is with the Mitre Corporation, Paper MP050000001, Feb 2005.
4 The Office of Safety and Mission Assurance, NASA Headquarters. *NASA Risk Informed Decision Making Handbook*. Version 1.0—NASA/SP-2010-576. 2010 Apr.
5 It should be noted that the RIDM has been adopted for security purposes by the IAEA and retitled as *Integrated Risk Informed Decision Management* (*IRIDM*). The publication is available in PDF format and is INSAG-25, available on the International Atomic Energy Agency website and other related websites: http://www-pub.iaea.org/books/IAEABooks/ 8577/A-Framework-for-an-Integrated-Risk-Informed-Decision-Making-Process. Accessed 2014 Oct 14.

6 The type of planning required may have to be prepared in conjunction with the environmental, safety, and health department personnel, the community services, and the local emergency planning committee. The response scenario should include anticipation of decontamination and transport of injured personnel, as well as community evacuation and notification where appropriate. Software such as ARCHIE, ALOHA, and other programs including detailed gas release and screening programs and where appropriate spill control and tracking programs and groundwater modeling programs should be used. These programs are available through the USEPA, the USCG, the NOAA, and the National Regulatory Commission. Some programs, especially those evaluating blast loadings, may be restricted to the United States or require special permissions for release. Commercial programs are available for 2D and 3D blast effects modeling on buildings. DNV also makes a program that combines chemical modeling with dispersion modeling and process hazard analysis, and it is suitable for a number of applications. The program is PHAST.

7 The security force must never be totally committed to a single event. Ripple attacks and diversionary events are a common factor used by aggressors and popularized by television drama. Consequently, the concept has entered the vernacular and tactics of the aggressor community.

8 Examples of the foregoing are as follows: (i) probabilistic risk would be a fence in a chemical facility on a remote island. It is reasonably secure from external attack. (ii) Actual risk would be a plant in a run-down, inner-city neighborhood where there is a lot of crime, a history of vandalism, and an organized and random robbery and theft. The risk factors are much much higher for some type of attack or break-in.

9 This site is found at Appendix I. Deepwater Horizon, http://docs.lib.noaa.gov/noaa_docu ments/NOAA_related_docs/oil_spills/BP_report/appendices_AA_Z/Appendix%20I. %20Deepwater%20Horizon%20Investigation%20Fault%20Trees.pdf, and in the Deepwater Horizon Accident Investigation Report, http://docs.lib.noaa.gov/noaa_documents/NOAA_ related_docs/oil_spills/BP_report/appendices_AA_Z/Appendix%20I.%20Deepwater%20 Horizon%20Investigation%20Fault%20Trees.pdf.

CHAPTER 3

ASSESSING TYPES OF ATTACKS AND THREATS WITH DATA SOURCES

The types of attacks and data sources that will be considered include external physical assaults, incidents, and accidents. We will start with the type of external physical assaults and attacks.

WEAPONS

One of the difficulties of protecting a facility from standoff weapons is the range of modern weapons and the damage they can do. The following is generally acknowledged as ranges of the various weapons commonly in use.

AK-47

Among the most popular weapons in use worldwide are the AK-47 (Kalashnikov) and its upgraded versions. The weapon is a 7.62 assault rifle that was in wide use in the Warsaw Pact and many African and Asian countries. It has the advantage of being rugged, and relatively cheap. The rifle files a 7.62×39 mm cartridge and is capable of selective fire, either single or automatic fire at a rate of 600 rounds/min with a muzzle velocity of around 715 m/s. The maximum range is 1000 m, but the effective range is practical at about 400 m unless it is equipped with telescopic sights. It has a feed system of 10–40-round box magazines. It also comes equipped with a folding stock. A later variant is the AK-74, which is currently used in the Russian Army. The weapon can be mounted with a rifle-prepared grenade that has a range of about 150 m. Most of the grenades are slightly over 1 lb (0.454 kg).

M16

The M16 is a US military-derived weapon. It uses a 5.56×45 mm NATO round, with a muzzle velocity of 948 m/s. It has an effective range of between 550 m (point target) and 800 m (area target). The rate of fire is between 12 and 15 rounds/min on sustained,

Industrial Security: Managing Security in the 21st Century, First Edition. David L. Russell and Pieter C. Arlow.

45 and 60 rounds/min on semiautomatic, and 700 and 950 rounds/min on cyclic fire settings. The M16 and its variants support a 40 mm grenade launcher, which has a range of about 150 m but which can be equipped with high-explosive and fragmentation shells. It is widely used in North and South Americas, the Middle East, India, Australia, and parts of the Southeast Asian peninsula including Korea, Vietnam, and Malaysia.

Sniper rifles

Sniper rifles come in a variety of sizes and applications. Depending upon the specific rifle, they are capable of hitting their targets accurately at distances from 350 m up to 1500 m. The exception to this is the 12.7×99 NATO and 12.7×108 Russian rifles, which are accurate up to 2000 m. The 14.5×114 mm (Russian) sniper rifle has a range of up to 2300 m. The obvious advantage of the rifle is its ability to hit a "head"-sized target at great distances, incapacitating or killing whatever it hits. Snipers generally record their longest shots, and it is not uncommon to find the longer ranges for sniper shots at 1250 m or more. Few attackers will use sniper rifles because of the cost. A .50 caliber sniper rifle can easily cost $15,000 or more, and the ammunition can be over $2.50 per cartridge.

Depending upon the type of attack, the AK-47 or the M16 is probably going to be most commonly available to an adversary. The anticipated range would be relatively short, generally under 50 m, and the volume of fire is often more important to an adversary rather than the accuracy of the fire.

MUZZLE ENERGIES FOR VARIOUS CARTRIDGES

The projectile fired by a rifle or a handgun, unless it is an explosive round, has a kinetic energy between 1000 and 4000 ft-lb of energy. Many handguns fall into smaller ranges. The kinetic energy delivered drops off somewhat significantly at ranges over 200 m, but that is dependent upon the type of projectile, its weight, and the initial velocity. Table 3.1 indicates the average initial muzzle energy for various types of projectiles commonly found in use.[1] Wikipedia gives the following information about projectile weapons.

The energy available for a projectile drops off with distance because air resistance slows the projectile. However, at a range of 200 m, the energy available from a rifle shell is between 40 and 70% of the initial muzzle energy. The higher impact energy at 200 yards seems to peak at around 70% when muzzle energy is about 6000 ft-lb.[2]

RIFLE GRENADES

The rifle grenade is generally an antipersonnel weapon that has a maximum range of around 150 m, with a fragmentation or other types of grenade. The fragmentation grenades have an effective shrapnel diameter of between 15 and 20 m. Modern

TABLE 3.1   Muzzle energies for various types of projectile weapons

Weapon type	Common designation	Muzzle energy (ft-lb)	Muzzle energy (J)
Pistol/rifle	22 long rifle bullet	117	159
Pistol	9 mm	383	519
Pistol	0.45 ACP	416	564
AK-47 and variants	7.62×39 mm	1,527	2,070
NATO standard round used by US and NATO forces	7.62×51 mm	2,802	3,799
Browning Machine Gun (NATO standard) for sniper rifles	12.7×99 mm	11,091	15,037
Heavy machine gun (Russian) antitank	57 caliber (14.5×114 mm)	23,744	32,000

[a]From www.wikepedia.org/wiki/Muzzle_energy.

shoulder-launched rifle-type grenades fire a 40 mm cartridge that has an effective range of about 150 m and a maximum range of 400 m. There are a variety of sources of public information on the types of grenades that can be fired from the various launchers, including thermobaric rounds for maximum effect over an area and high-explosive rounds for antitank use. Any of these weapons used against an industrial target would cause a large amount of damage and destruction.

ROCKET-PROPELLED GRENADES AND MORTARS

Rocket-propelled grenades (RPG) are a shoulder-mounted weapon that has a variety of warheads. The warheads can be either high explosive or shaped charge or thermobaric charges. A typical RPG as described by Wikipedia has the following properties:

> It is 40–105 millimeters in diameter and weighs between 2.5 and 4.5 kilograms. It is launched by a gunpowder booster charge, giving it an initial speed of 115 meters per second, and creating a cloud of light grey-blue smoke. The rocket motor [2] ignites after 10 meters and sustains flight out to 500 meters at a maximum velocity of 295 meters per second. The grenade is stabilized by two sets of fins that deploy in-flight: one large set on the stabilizer pipe to maintain direction and a smaller front set to induce rotation. The grenade can fly up to 1,100 meters; the fuse sets the maximum range, usually 920 meters.

Fortunately, the accuracy of the weapon is relatively poor. The same article cited earlier suggests that at distances beyond 180 m, the accuracy decreases with increasing distance, especially on a moving target.[3]

Modern mortars are high-arc weapons. They are tube fired and muzzle loaded. The diameter and size of the mortar vary between 60 and 120 mm, but there are a number of homemade mortars that are also in use. Because the mortar is fired from the ground in a high arc, it is a plunging weapon, coming into the target from overhead.

The shell fired by a mortar can, depending upon the type of mortar, contain as much as 200 kg of explosives. Larger mortars are usually motor mounted, and the difficulty of carrying multiple high-explosive shells, especially the larger rounds, can limit their applicability. A typical field mortar is the British 81 mm mortar. The system is relatively light in its original configuration; it weighs about 41 kg, but a lighter version is now available. The mortar is crewed by a team of three people and has a range of under 100 m to almost 5900 m. The payload of the mortar is several kilograms of high explosive, but exact information on the payload is not readily available. The accuracy is very good, and at several hundred meters, an experienced crew can drop the mortar round inside a 10 m circle on a consistent basis.

EXPLOSIVE ENERGIES

The amount of energy available in various types of explosions naturally varies with the explosive compound and the manner in which it is delivered. Explosives are measured in the equivalencies to trinitrotoluene (TNT). One gram of TNT has an energy release of 4162 kcal, although it is often considered as having energy levels between 4100 and 4600 kcal/g. The 4162 kcal/g is a definition roughly equivalent to 23,118,800 ft-lb of energy—per pound of TNT. All explosives are rated in comparison with TNT. Table 3.2 gives a rough conversion for common explosives.

TABLE 3.2 Energies of various explosive compounds

Explosive	Speed of detonation (m/s)	Relative effectiveness with comparison to TNT
Ammonium nitrate	2250	0.45
Ammonium nitrate/fuel oil	5570	0.8
Black powder	800	0.55
Composition B (63% RDX, 36% TNT) (military explosive)	7840	1.35
Gelatin (92% nitroglycerine (NG), 7% nitrocellulose; military explosive)	7970	1.60
RDX (hexogen)	8700	1.60
Nitrocellulose (13.5% N)	6400	1.10
Nitroguanidine	6750	1.00
NG	6750	1.00
75% NG dynamite	8120	1.25
PETN	8400	1.66
Tetryl	7770	1.25
Pentolite (56% PETN, 44% TNT)	7520	1.33
Semtex 1A (76% PETN, 4.7 RDX)	7670	1.35
Tetryl	7770	1.25

[a]Wikipedia—Table of explosive detonation velocities.

Impact of explosives

The effectiveness of the explosive is in the shock wave it creates. The shock wave creates a rapidly expanding spherical overpressure on all that surrounds it. As the wave expands, it slows and dissipates. The wave generally moves at the speed of sound, and as the wave passes, it can generate a negative pressure wave as the displaced air returns after the shock has passed. This is shown in Figure 3.1.

The timing of the pressure wave is in milliseconds, and the velocity depends upon the power and brisance of the explosive as shown below.

The points for t_A and t_d are influenced by the amount of explosive material and the speed of sound. The calculation of blast overpressures can be approximately scaled by the following formula:

$$Z = \frac{R}{W^{1/3}}$$

where Z is a scaled distance, R is the actual effective distance from the explosion, and W is the weight of the explosion in TNT equivalent kilogram. The scaled distance is then used to estimate the overpressure by the following formulas:

$$P_{(\text{In BARs})} = 6784\frac{W}{R^3} + 93\left(\frac{W}{R^3}\right)^{1/2} \qquad \text{where } P \text{ is in bar.}$$

$$P_{(\text{kPa})} = \frac{1772}{Z^3} - \frac{114}{Z^2} + \frac{108}{Z} \qquad \text{where } P \text{ is in kPa.}$$

If the pressure wave encounters a structure, it will be reflected off the front face creating a localized reflected overpressure. The pressure wave will continue until it encounters the back side of the building where it will generate a negative pressure wave of the appropriate magnitude on the structure.

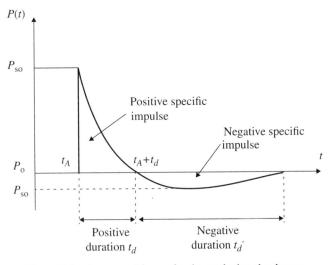

Figure 3.1   Power and forces for the explosive shockwave.

TABLE 3.3   Damage rates from a 3 to 5 m/s explosion

Organ	Pressure (kPa)	Damage
Ears	6.2 at 0.207 kPa-m/s	Threshold shift in hearing (log scale)
	2.0 at 0.069 kPa-m/s	
	34.5	Threshold eardrum rupture
	103.4	50% eardrum rupture
Lungs	206–276	Threshold for lung damage
	$P > 552$	50% lethality for lung damage
Whole body	670–830	Threshold for whole body
	900–1250	50% lethality
	1380–1750	<1% survival

[a]From Blast injuries. Estimated human tolerances for single, sharp, rising blast waves. Courtesy of Bowen TE and Bellamy RF, editors. *Emergency War Surgery*. Washington, DC: United States Government Printing Office, 1988. http://emedicine.medscape.com/article/822587-overview.

TABLE 3.4   Explosive pressures from a 1500 kg ANFO explosion

R (m)	20	50	100	150	200	250	300
Z		1.89	4.72	9.43	14.14	18.86	23.58	28.3
Overpressure (kPa)	290	34.67	12.28	7.70	6.67	4.51	3.75

Sufficient overpressure can destroy nonhardened buildings, but the real damage can be to the personnel in the blast area. Loss of life and limb and permanent hearing loss can result. According to the US Department of Defense's *Structures to Resist the Effects of Accidental Explosions*, UFC 3-340-02 (December 2008), the range of damage to humans starts at about 5 psi for hearing loss, and the damage increases substantially to about 80 psi where the lungs collapse. Table 3.3 helps explain the damage.

Example of an explosive device:

A truck with an explosive device estimated a 1.5 metric tons is approaching a checkpoint. If we assume that the material is ANFO, how far out is the blast range for buildings and personnel?

First, calculate the power of the explosive in TNT equivalent.

ANFO is equivalent to about 0.8 TNT, so 1.5 MT of ANFO is equivalent to 1.2 MT of TNT.

Note that the TNT will have a density of about 1.5 kg/l so the volume will be just under a cubic meter—actually (0.8 m^3).

For distances, the scaling distance Z will be the measured distance $Z = R/(1200)^{1/3}$ or for this explosive quantity $Z = R/10.627$.

Then, the pressures in kPa are in Table 3.4.

A plot of the data is shown in Figure 3.2.

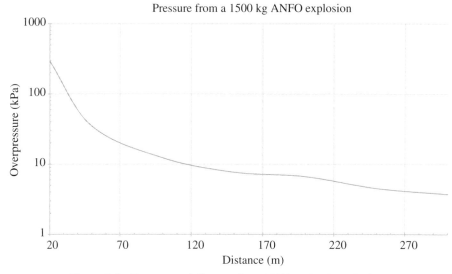

Figure 3.2   Pressure and distance for a 1500 kg ANFO explosion.

According to these calculations, anyone closer than about 30 m from the source will have about a 1% chance of survival, while severe injuries will occur at 20–30 m, and people at about 100 m will have a 50% chance of eardrum rupture. Glass breakage occurs at about 15–20 kPa. That suggests that a distance of 100 m should be sufficient to protect most of the command and control facilities of a modern industrial plant and dispatch center should be at least 100 m from a checkpoint unless the glass in the center is blast proof and reinforced.

This also suggests that any shipping and receiving inspection stations should be in a remote location, away from the warehouse and well away from the center of the plant.

OTHER TYPES OF INCIDENTS AND ACCIDENTS

For an examination of the likelihood of other chemical accidents or incidents, it is recommended that one should use a chemical hazard incident evaluation program. There are several programs on the market and a free program from NASA/USCG, which is called ALOHA. Other programs are available from the Protective Design Center (US Department of Defense) that has proprietary software, the US Nuclear Regulatory Commission that has chemical specific hazard spreadsheets in the public domain, and a number of commercial software programs.

There are also a number of good references on the subject of blasts, explosives, and building hardening. These include:

US Department of Energy. A manual for the prediction of blast and fragment loadings on structures. Amarillo (TX): U.S. Department of Energy; 1992. Report no. DOE/TIC 11268.

Air Force Engineering and Services Center. Protective construction design manual. Tyndall Air Force Base (FL): Air Force Engineering and Services Center, Engineering and Services Laboratory; 1989. Report no. ESL, TR87-57.

Committee on Feasibility of Applying Blast-Mitigating Technologies and Design Methodologies from Military Facilities to Civilian Buildings, Division on Engineering and Physical Sciences, Commission on Engineering and Technical Systems, National Research Council. National Academies Press, Oct 26, 1995, Political Science, p. 112.

Ngo T, Mendis P, Gupta A, Ramsay J. Blast loading and blast effects on structures—an overview. Electric Journal of Structural Engineering Special Issue Loading on Structures 2007;75–91.

NOTES

1 The 5.56×45 mm, the 7.62×39 mm, and the 7.62×51 mm are usually considered to be military rounds and might probably be in the hands of an attacker unless he/she can gain access to 50 caliber weapons.

2 An examination of published data for rifle shells listed on http://www.chuckhawks.com/rifle_ballistics_table.htm indicates that the energy delivered generally follows a fourth-order curve with a peak of about 70% somewhere around 6000 ft-lb of muzzle energy. A rough correlation coefficient indicates that the R^2 coefficient is 0.754 for the sixth power curve.

3 A US Army evaluation of the weapon gave the hit probabilities on a 5 m wide (15 ft), 2.5 m tall (7.5 ft) panel moving sideways at 4 m/s (9 miles/h) [8]. This probability decreases when firing in a crosswind due to the unusual behavior of the round; in a 7 mile (11 km) per hour wind, the gunner cannot expect to get a first-round hit more than 50% of the time beyond 180 m. (Source: *TRADOC Bulletin 3, Soviet RPG-7 Antitank Grenade Launcher*. United States Army Training and Doctrine Command, Monroe, VA. November 1976.)

CHAPTER 4

EVALUATING A COMPANY'S PROTECTIVE SYSTEMS

SURVEYS AND ASSESSMENTS

The evaluation of a company's protective systems involves a survey, an assessment of resources, planning, and a lot of work and imagination. The first item is the survey. The survey starts with a detailed mapping. In the event that it is a new facility, one needs to incorporate safety, security, and environmental response to catastrophes into the designs. Failure to plan for the worst will lead to the worst happening, and it never occurs at a convenient time.

There are several considerations that can be key to security. Site layout and physical arrangement are probably the most important consideration. That includes natural and artificial barriers (which may make defense a bit easier), physical intrusion barriers such as fencing, protective systems, and alarms. Physical arrangement of the plant and its environs can also be key in mitigating natural disasters such as hurricanes, tsunamis, tornadoes, and other extreme weather events, including extreme precipitation.

Another security consideration is logistical and supply chain operations, which include shipping, receiving, and such items as supply lines, and remote locations such as pumping stations. The assessment of this element should also consider the security of the sources of supply as well as the security of the transportation to the plant. One often-overlooked critical need for a plant is the knowledge of the security, source, and location of the cooling water supply, whether it be from the ground or from river, the sea, desalinated water, or even treated and recycled wastewater.

Emergency planning is also a part of plant security. While emergency planning response is often thought of as a part of contingency planning, security has a special role to play in emergency situations, including restricting unauthorized personnel, permitting emergency and service personnel, traffic control, and communications and notification of potentially affected employees.

SITE SECURITY ASSESSMENTS

Some of the previously mentioned concerns are self-evident. If the site is located on a peninsula or has a substantial water frontage for shipping, the security of the waterfront and the shipping will depend upon the type of operations, and the amount of investment in security, both manpower and facilities.

Checklists

The best way to conduct a preliminary security survey is to develop and use a checklist for each section or area of the facility. This may require substantial work and division into several subsections, such as electrical, communications, transportation, supply chain, physical security, and cyber security. Process hazard analysis, spill response, and emergency response procedures should be included or considered in the overall security plan. Each of these areas deserves a separate security plan that addresses the need for security, accident prevention, and an analysis of the deficiencies, in these areas, and, most importantly, *recommendations for repairing the deficiencies*. One of the things that the security plan does not address is the structural integrity of the building in response to attacks by weapons or explosives. While it is fondly hoped that there will never be such an attack, prudent wisdom suggests that at least some of the buildings, such as the communications center and the guardhouse, should be not only bulletproof but blast resistant as well.

Checklists can be detailed or general. Some at the lowest level of detail merely ask if there is a plan of the type in place. The more detailed checklists may go into the details of the operations, and when it comes to computer security, the checklists abound. Every major manufacturer of computer equipment has its own set of checklist recommendations for protecting the plant's operating systems and preventing data breaches or cyber intrusion.

The following checklist is a place to start with physical security and plant borders, but one may want to add additional criteria as the facility may require. The checklist does not address the possibility of explosives or bombs, except in passing, and it is not designed for those purposes. It is just one example of the type of information and the level of detail that should be dealt with in good physical security.

We have provided two examples of checklists in the appendix of this book to illustrate the difference in detail. The first checklist is for physical security and has been adapted from the US Department of Agriculture (USDA) Checklist for Physical Security. The checklist was adapted for plant operations, whereas the USDA checklist is primarily detailed for a single building.

Cyber security checklist

A detailed Cyber Security Checklist© prepared by the US Cyber Security Consequences Unit is 42 pages, prepared in 2007, and can be found at the following address: http://www.usccu.us/documents/US-CCU%20Cyber-Security%20Check%20 List%202007.pdf.

The second checklist is a part of a much longer document prepared by www.Bereadyutah.gov, an arm of the UTAH State Government, and a portion of the 12-page document that deals with the elements of cyber security is reproduced in the appendix. The appendix does not consider the physical security aspects such as badges and personnel and contractor investigation and control that are addressed in the larger document, but this excerpt is being presented as an example of the type and level of detail a good security plan has to consider.

In reviewing the checklists, it is good to note that there are a number of programs that support the checklist. It is also a good idea to pay attention to various security standards, both national and international for guidance. If transportation and operational security is of concern, the Model-Based Risk Analysis Program or the CARVER + Shock Programs can be of help in planning and identifying potential security problems.

LIGHTING

Critical to security is lighting. Perhaps two of the best guides to facility lighting are the *Outdoor Lighting Code Handbook* by the International Dark-Sky Association (http://www.darkskysociety.org/handouts/idacodehandbook.pdf) and the more general International Commission on Illumination (CIE) Technical Report *Road Transport Lighting for Developing Countries*, which is found at http://files.cie.co.at/180.pdf. Some of the material in both guides is duplicates, but the CIE has a more thorough discussion of transportation issues and illumination. The Dark-Sky Association (DSA) handbook may be more useful in industrial applications. The US Navy also has an excellent manual on lighting.

Some of the common recommendations and observations from both include a discussion on the appropriate lighting levels for visibility. The DSA and CIE define various levels of lighting as the following:

- Zone E1: Areas with intrinsically dark landscapes. Examples are national parks, areas of outstanding natural beauty, areas surrounding major astronomical observatories (but outside Zone E1A—see below), or residential areas where inhabitants have expressed a strong desire that all light trespass be strictly limited.
- Zone E2: Areas of low ambient brightness. These are suburban and rural residential areas.
- Zone E3: Areas of medium ambient brightness. These will generally be urban residential areas.
- Zone E4: Areas of high ambient brightness. Normally, these are urban areas that have both residential and commercial use and experience high levels of nighttime activity.
- Zone E1A: Dark-Sky Preserves.

For most applications in industry, nighttime illumination will fall somewhere in the Zone 3 or Zone 4 categories. Illumination around the perimeter for control should be at least a Zone 3 area, and areas around checkpoints should be Zone 4. All external illumination should be directed outward and downward at a 45° angle so as to be in the eyes of any intruder or anyone approaching the facility.

Table 4.1 is taken from the US Army Field Manual for lighting and has been condensed for clarity.

Lighting should be limited to those areas that need it or should be protective. For example, glare lighting for approaching a guard location is designed to conceal the actions and location of the guards, while fence lighting should be in relatively narrow strips to focus on intrusion detection while providing dark areas for patrols.

Additional lighting guidance is provided by the *UK's Centre for the Protection of National Infrastructure*:

- Lighting can be an important security measure but may in fact assist an intruder if used incorrectly.

- The purpose of a lighting system should include the following: (i) deter intrusion, (ii) reduce intruder's freedom of action, (iii) assist in the detection of intruders either by direct observation or by closed-circuit television (CCTV), and (iv) provide concealment for guards and patrols.

- It is often difficult to arrange lighting so that it achieves the desired ends, and compromise is often required to balance those needs.

- Lighting should be coordinated with the CCTV requirements to create a lighting environment that does not illuminate guards or patrols and that will support them.

- The illumination should be balanced so that there are gradual increases between brightly illuminated areas such as roads, and less well lit areas such as parts of the tank farm or remote areas such as in the rail yard. The purpose is to prevent dark spots where an intruder can find concealment.

- Lighting columns should not provide the intruder with an aid to scaling the fence.

TABLE 4.1   US army field table for lighting security

Location	Type of lighting	Width of strip (m)		Lumens at ground level
		Inside fence	Outside fence	
Perimeter of outer area				
Isolated	Glare	8	75	1.7–2.0
Isolated perimeter	Controlled	3	20	5
Vehicular entrance	Controlled	15	15	11–15
Pedestrian entrance	Controlled	8	8	22–25
Railroad entrance	Controlled	8	8	11–15
Open yards	Controlled			20–25
Vital structures	Controlled	8–10 m from structure		11–15
Deck or pier	Controlled			11–15

- Floodlighting should be used to illuminate building exteriors and entrance-ways to silhouette the intruder, and that should be coupled with CCTV so that the intruder is easily identified and is not in shadow. It is equally important to provide shielded lighting that will illuminate the intruder's face as well as his/her profile. Normally, lights are mounted out of intruder's reach, but in the area around entrances, judicious use of CCTV should be coupled with face lighting.

- In some cases, this may require positioning of minicameras at heights that will enable facial recognition. These minicams are rugged, are weatherproof, and have been installed in a number of banks at teller windows. They are coupled to the area CCTV system and enable facial recognition.

- Lighting at gatehouses and perimeter entrances should be bright enough to reveal approaching vehicles and pedestrians and allow guards to identify them, verify passes, carry out vehicle searches, and conceal guards within the gatehouse while allowing them to see out.

PERIMETER BARRIERS: DESIGN NOTES AND COMMENTS

The perimeter barrier has been partially addressed in some of the material earlier. This chapter will provide some reasons for the various recommendations as well as practical guidance on setting up perimeter barriers. There are a number of opinions as to the correct height for a perimeter barrier, and those numbers range from about 2 to 5 m. Regardless of the height, the perimeter barrier should be difficult to scale with a ladder or climbing tools, and it should be topped with barbed wire and at least one coil of razor wire (barbed tape) as shown in the following.

If space permits, dual fencing should be used. There should be a clear zone of at least 6 m outside the outer fence and at least 6 m between the inner and outer fences and preferably another 6 m between the inner fence and any buildings. This area must be cleared and maintained so that concealment in any of the zones is not possible. The fencing must be designed so that it is not convenient to place a ladder or scaling device close to the barrier. Some fencing systems use several coils of razor wire on the outside of the fence at its base to discourage climbing.

Additional security can be gained by using various types of thorny bushes or trees. These include barberry, holly, hawthorn, pyracantha, locust, quince, wild rose, blackberry, and cholla. There is an excellent brief discussion on barrier plants on the Internet at http://thesustainablelife.tumblr.com/post/5864752175/home-security-the-way-nature-intended.

Fence posts should be a minimum of 3″ (7.5 cm) diameter and preferably 4″ (10 cm) diameter where there is the slightest possibility of vehicle impact. The fence itself should have two layers of steel cables run through the fence, and the spacing between the fencing should be supported with diagonal guys on turnbuckles. The fencing should be imbedded in concrete, and that will be illustrated in Figures 4.1, 4.2, and 4.3.

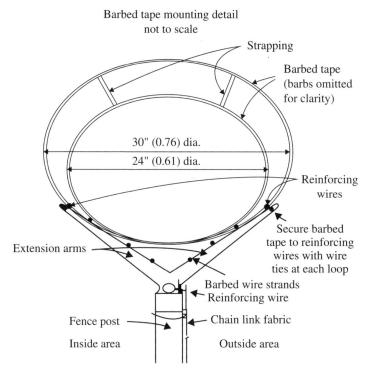

Barbed tape mounting detail
not to scale

Strapping

Barbed tape
(barbs omitted
for clarity)

30" (0.76) dia.

24" (0.61) dia.

Reinforcing
wires

Secure barbed
tape to reinforcing
wires with wire
ties at each loop

Extension arms

Barbed wire strands
Reinforcing wire

Fence post

Chain link fabric

Inside area

Outside area

Figure 4.1   Detail for top of protective fencing. From Explosive Blast. http://www.fema.gov/media-library-data/20130726-1455-20490-7465/fema426_ch4.pdf.

If the fencing is to provide privacy as well, it should have metal slats inserted through the fence openings. The use of the metal slats will provide some protection from prying eyes, but the slats act as a wind barrier and that will dramatically increase the wind loading on the fencing, and if there is a sensor on the fencing to measure vibration, it will be virtually useless in almost any wind greater than about 5 m/s (18 km/h or 11 miles/h).

It is also important that the fencing withstand the force of the strongest wind without damage and still be able to act as a potential barricade against vehicle intrusion. The vehicle intrusion depends upon the force and height of the vehicle and the bracing of the fence. The following calculations will be of some help in determining wind loadings and should be applicable to vehicle impact loading as well.

For routine wind loading, there are two essential components, a moment (M) (bending force on the fence post where it meets the foundation) and a horizontal force (F) in shear against the footings for the fence. There is also another factor known as porosity of the fence (h), which is the porosity of the fence ((total area of fence − cross-sectional area of the wire, bars, posts, and any slats inserted)/(total area of fence)).[1]

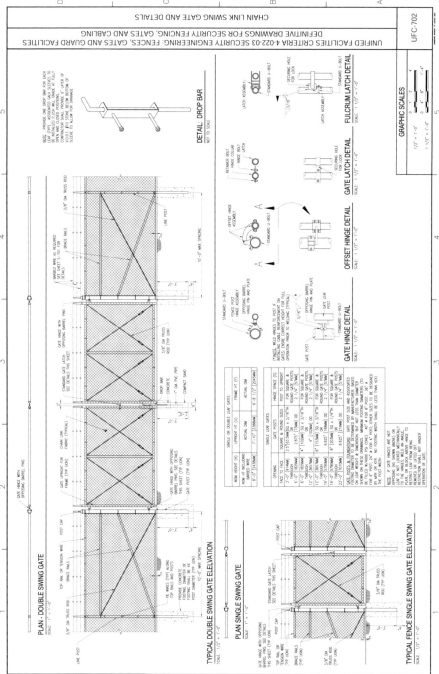

Figure 4.2 Security fence detail—elevation. From US Army Manual on Physical Security UFC 4-22-03.

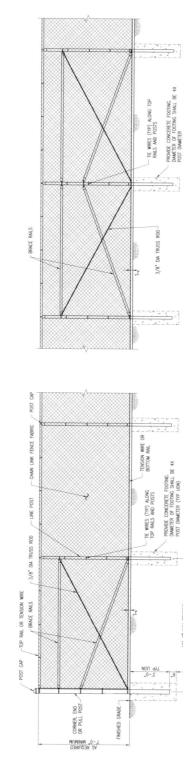

Figure 4.3 Additional details on security fencing. The drawings were reproduced from the US Army Manual UFC-4-22-03.

The two formulas are

$$F = \left[\frac{2.12}{1 + \left(\dfrac{\eta - 0.24}{2.07} \right)^2} - 1.72 \right] h\rho U^2$$

$$M = \left\{ 0.4 \exp\left[-0.5\left(\frac{\eta - 0.2}{0.81} \right)^2 \right] - 0.16 \right\} h^2 \rho U^2$$

where h is the porosity of the fence, U is the wind speed in meters/second, h is the height of the fence in meters, and ρ is the density of the air. From this equation, the engineering department can calculate the reactions on the foundations for the posts and decide how deep and how large to make them. Note that this does not guarantee that the fencing can withstand or stop a vehicle ramming the fence, but it will slow it down.

An alternative design method is supplied by the Allan Block Company. In their alternative design, they relate wind speed and pressure and then combine that with exposure conditions to determine the pressure on the fence. The pressure graphs are shown in Figure 4.4.

For reference purposes, 1 km/h is equivalent to 0.62121 miles/h and 1.0 kPa is equivalent to 0.04788 lb/ft². Table 4.2 is further modified by factors that depend upon exposure and height, as shown below.

The pressure diagram for computation of the fencing overturning moment is triangular with the highest forces at the top of the fence or wall.[2]

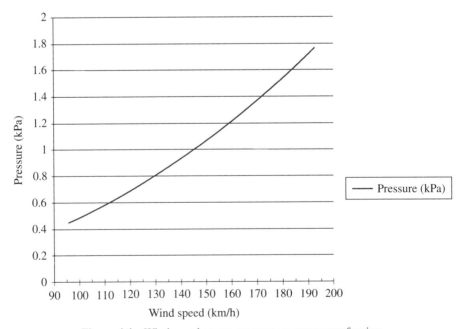

Figure 4.4   Wind speed versus pressure on nonporous fencing.

TABLE 4.2   Pressure coefficients for nonporous fencing

Exposure condition	Coefficient for fences <3.7 m (12 ft)	Coefficients for fences >3.7 m (12 ft)
Urban or wooded exposure with trees over 10 m within 100 m of the site	0.68	0.85
Unobstructed exposure with few trees over 10 m within 50 m of the fence	0.9	1.2
Unobstructed exposure similar to barren or open desert or open waters	1.25	1.5

Depending upon the location and the design of the fence, it may be necessary to embed the bottom of the fence below ground or at least provide secure anchors so that it cannot be easily lifted for an intruder to slide under. Optionally, the fence should be equipped with fence shaker detectors; there are several manufacturers of this type of detector available, and the usual design is to install a shaker sensor every 3 m. On windy nights, however, the shaker detector can generate false positives unless the sensor system is calibrated to the natural frequency of the fence/wind speed combination.

Another type of sensor to be used in the area between the primary and secondary fences is an induction sensor. The sensor relies upon a magnetic field set up by two buried wires about 2 m apart. Anything crossing between those two wires will change the electromagnetic (EM) signature of the wires, and the voltage will change. The change in the EM profile of the wires will help to locate an intruder, by defining the distance along the sensor net.

Additional security can be had when radar (or a microwave detection system) or other trip light detection system is installed. Some of the detectors use lasers and the ranging on the target can be quite accurate. Other detectors include UV or infrared (IR) light beams to be interrupted to form or create an alarm signal. These line-of-sight detection systems will require the pathways to be kept clear. Depending upon the type of sensor, it may be able to range the exact location of the intruder from the source. Some of the detector beams have difficulty with flying birds and crawling insects blocking the light paths, causing false alarm signals. Multiple redundant paths can minimize this problem, and even with lasers, that can include beam splitting so that both light paths have to be interrupted before an alarm signal is generated. Concealment or camouflage of the transmitter and receivers is also necessary for effective security.

CCTV

CCTV can be a vital part of any security system. In order for it to be effective, especially on perimeter systems, it must have a clear line of sight and should include thermal image detection as well as visible light detection. Recent advances in CCTV imaging now permit high-definition images to be captured. There are a number of very good guides to the selection of CCTV on the Internet. A few of those include:

http://www.boschsecurity.us/NR/rdonlyres/1A4F9B44-0376-4FC8-A735-151F02021082/0/SelectingtheRightCCTVCamera.pdf by Bosch. The guide is excellent and raises several very good points about the range of the camera and the lenses.

http://www.inter-pacific.com/documents/education/Inter-Pacific_Camera_Selection_Guide.pdf is from Inter-Pacific. The guide is also excellent, but does not provide as much information and detail as the Bosch guide.

http://www.iviewcameras.co.uk/pdf/cctv_buying_guide.pdf is a compact and simple guide for introduction to CCTV. It does not provide as much information as either guide listed above.

http://promaxusa.com/cctv-lens-chart.html provides graphical illustrations of the resolution and color for various types of camera lenses.

The principal considerations in the selection of a CCTV service relate to resolutions (no. of lines per centimeter or inch) where a higher number is better; the responsiveness of the system to varying light levels, which can be up to 107,000 lx in bright sunlight and 0.001 lx on an overcast night; the responsiveness of the camera to IR light or ability to "see" at night; whether or not the camera is to be using color imaging; the ability of the camera/lens system to compensate for backlight; the focal length of the lens in the camera and lens aperture system (lower numbers are better than high numbers); the field of view of the camera; the ability to zoom to focus on distant or nearby objects; the camera enclosure; the ability of the camera system to scan an area; and the signal-to-noise ratio for the camera (signal-to-noise ratio should be at least 40 dB or about 100:1 or better for clear pictures). Higher ratios provide better pictures.

Black-and-white video costs less than color, but each has their advantages. For night work, IR-capable cameras work better with black-and-white video systems and black-and-white systems are cheaper than color systems. The video system should be coordinated with the perimeter and other lighting in the facility. Especially for perimeter lighting, some type of remote control to increase lighting levels if an alarm is triggered may be warranted.

The physical layout of the plant and its perimeter will, to an extent, influence the type of security system you are going to use. If you are concerned only with the perimeter, some type of pole-mounted cameras that show the perimeter areas probably will suffice, but there are considerations about the cabling for the CCTV system. Coaxial cabling tends to be significantly more expensive than twisted pair cabling, and the coaxial cabling does not allow for sound. If the distance between the cameras and the receiving station is great, consider either a microwave system or plan on installing signal boosters in the lines. Cameras should have highest resolution possible, should compensate for backlight, glare, and should have a sampling rate and shutters fast enough to read license plates on moving objects. Power is a consideration as well. The cameras should be supplied with standby power, and that will be either 24 or 12 V, and the power supply should be large enough to provide for the camera motors so that the scanning and autofocus features are not interrupted.

For long transmission distances, repeaters or signal amplifiers may be required, and these can represent a potential nuisance or vulnerability if they are not properly protected and the junction box housing designed to be tamper and weatherproof. If microwave transmission is used, the frequencies should be 5.6 GHz, and if possible, frequency hopping and signal encryption should be employed.

It is vital that a second central dispatch center, probably at some distance from the front gate, should be established and that the CCTV signals should be routed to the backup facility as well as to the primary facility. In the event that either guard dispatch or control center is taken out in an attack or an accident, a fully functioning backup system should be available to respond to emergencies.

WINDOWS AND DOORS

Perimeter windows and doors should be alarmed, depending upon the contents of the building, the use of the doors, and the occupancy of the building. Thus, a building on the perimeter should have easily accessible doors that face the perimeter alarmed with coded entry pads.

Doors on the lower floors should have dead bolt locks and metal frames. Door hinges should be accessible only from inside, and a minimum of three hinges are recommended. Doors should be heavy gauge metal, and the locks should be dead bolt locks wherever possible. Fire doors that have panic hardware for quick release should be equipped with an alarm.

Windows facing the perimeter or where there is a possibility of forced entry should have metal frames that are integral with the wall and tied into the building wall structural materials. Windows in facilities where there is the slightest possibility of an explosion or overpressure should be coated with a fragment retention window film that is bonded to the window frame. These films are made by a number of manufacturers including 3M. The lower floor windows should have alarms and bars on the windows sufficient to prevent forced entry. Those windows should open inward rather than outward.

NOTES

1 Dong Z, Mu Q, et al. An analysis of drag force and moment for upright porous wind fences. J Geophys Res 2008;113:D04103. doi: 10.1029/2007/JD009138.
2 Source: Allan Block Fence Engineering Manual, available from the http://www.allanblock. com/literature/PDF/FenceEngineeringManual.pdf. Accessed 2014 Oct 14.

CHAPTER 5

PORT SECURITY

Port security is a separate chapter because the subject material covers physical security plus the requirement to protect large, mobile, high-value targets, plus fixed-port facilities. Ports often have the added disadvantage of trying to eliminate threats while discriminating between a potential attacker and legitimate commercial vessels or civilian watercraft. Swimmers and submersibles can also pose threats to the port and the ships therein. Port security is different because the high-value targets are not continuously at the dock, but substantial damage to the loading and unloading facilities can impair the function of the operations that the dock supports. The level of security needs to increase with the arrival of ships and the likelihood of an attack, but a certain minimum standard of security needs to be maintained at all times.

RANKING THREATS

The organization *Maritime Security Outlook*[1] has suggested that the principal risks associated with a port include the following threats:

Threat   **Rank of threat 1–5, 5 being the most likely**

Natural threats

    Hurricane

    Tornado

    Flooding

    Tsunami

    Earthquake

Man-made/accidental threats

    Hazardous material (HAZMAT) spill

    Fire

Industrial Security: Managing Security in the 21st Century, First Edition. David L. Russell and Pieter C. Arlow.
© 2015 John Wiley & Sons, Inc. Published 2015 by John Wiley & Sons, Inc.

Extended power outage

Transportation network loss

Operator error

Loss of data center/networks/IT infrastructure

Intentional acts—delivery vectors

Container

Boat

Cars/trucks

Swimmer

Military grade submarines

Small, radar-dodging, self-propelled semi-submersibles (SPSSs)

Disgruntled individual authorized to be on the property

A criminal not authorized to be on property

Weapon threats

Chemical, biological, radiological, and nuclear (CBRN)

Enhanced explosives/improvised explosive devices

Conventional weapons

Cyber attack

LEVELS OF PORT SECURITY

The US Coast Guard (USCG) has studied the issue of port security and has classified the various levels of port security into three levels and has developed a set of minimum recommendations around those levels. The guidance document has been coordinated with the International Maritime Organization (IMO). These levels are as follows:

- Level I—The degree of security precautions to take when the threat of an unlawful act against a vessel or terminal is, though possible, not likely
- Level II—The degree of security precautions to take when the threat of an unlawful act against a vessel or terminal is possible and intelligence indicates that terrorists are likely to be active within a specific area or against a type of vessel or terminal
- Level III—The degree of security precautions to take when the threat of an unlawful act against a vessel or terminal is probable or imminent and intelligence indicates that terrorists have chosen specific targets

These guidelines are intended for application to all US waterfront facilities. The guidelines and regulations for port security are described in detail in Chapter 33 of the US Code of Federal Regulations, Parts 125–128 and 154; NIVC 3–96, NVIC 1–97, and IMO Circular 443.

Security response plans

The foundation of all three levels of port security starts with a plan for security and response (S&R plan). The minimum elements of the S&R plan should include the following:

- Identification procedures
- Access control procedures
- Internal security requirements
- Designated restricted areas
- Perimeter security plan
- Security light and maintenance plan
- Security alarms, video surveillance, and communications systems
- A name designated security officer (with contact information)
- Training program for the security force
- Training for employee security awareness
- Security communications, including prearranged agreements with the local police and fire
- Set procedures for upgrading security to Level II and Level III

Recommended procedures

- A set of detailed response procedures for the following scenarios:
  - Unauthorized personnel discovered at the facility
  - Unauthorized or illegally parked or abandoned vehicles in or near the facility
  - Unauthorized vessels moored along the waterfront property
  - Bomb threat
  - Suspicious persons or activity response
  - Mail handling
  - Unknown or suspicious package discovery and response

The USCG has developed guidance for scenario development for vessels based upon a threat matrix table as discussed in the previous chapters. This threat matrix is to be used to develop the vessel security plan. This threat matrix is documented in the *USCG Navigation and Vessel Inspection Circular 10-2, entitled Security Measures for Vessels, dated October 21, 2002*. The document is available on the Internet at http://www.uscg.mil/hq/cg5/nvic/pdf/2002/10-02.pdf (Appendix B). The minimum scenarios that the Coast Guard mandates for the vessel security plan include those types of attacks that could damage or destroy the vessel (either from contact or from a distance using standoff weapons), sabotage the vessel, control the vessel, create a pollution or toxic material incident under a variety of circumstances, or use the vessel for illegal transport of personnel or weapons.

The following is suggested by the USCG and IMO as being suitable at a minimum for a good security plan. The USCG also strongly suggests that the plan is addressed in the specified order.

IDENTIFICATION PROCEDURES FOR PERSONNEL SCREENING

All persons entering a facility should possess and show a valid photo ID card to gain facility access. Individuals arriving by motorcycle should remove helmets to assist in identification. Security personnel or competent authority should verify that ID card matches the person presenting it. While conducting roving patrols, security personnel or competent authority should challenge unknown or suspicious personnel to identify themselves with a photo ID card. At passenger terminals, security personnel or competent authority should refer to their terminal security plan for procedures for allowing personnel into the secure area of the passenger terminal. The following specifics apply.

Employees

All employees should be required to show an employee/union photo ID prior to entry. Facilities that do not use employee/union photo IDs should cross-reference employee's identification with employer-supplied access lists. While on the facility, personnel should possess valid identification and present upon request by security/ government representative. The facility should have a verification process to ensure employees entering have valid business.

Vendors/contractors/vessel pilots

All vendors/contractors/vessel pilots should be required to show a valid photo ID prior to entry. While on the facility, all vendors/contractors/vessel pilots should possess valid ID and present upon request by security/competent authority/government representative. Vendors, contractors, and pilots visits should be scheduled in advance. If the arrival of vendors, contractors, or pilots is not prearranged, entry should be prohibited until their need to enter is verified by proper authority (as identified in the facility security plan).

The facility should have a verification process to ensure vendors/contractors/ vessel pilots entering have valid business. The use of access lists that preauthorize regular contractors, vendors, and pilots to enter the facility or board vessels moored at that facility is permitted in lieu of the daily schedule requirements.

Truck drivers/passengers

All truck drivers (for cargo) and passengers (when allowed by the facility) should be required to show a valid photo ID prior to entry. While on the facility, truck drivers/passengers should present this ID when requested by security/government representative. The facility should have a verification process to ensure drivers entering have valid business (e.g., checking booking numbers).

Visitors (all personnel not falling into other categories)

All visitors should be required to show a photo ID prior to entry. While on the facility, all visitors should be required to have photo ID and present it upon request by security government representative. Visitors should be scheduled in advance. If not, entry should be prohibited until proper authority authorizes visit (as identified in the facility security plan). The facility should have a verification process to ensure visitors entering have a valid purpose for their visit.

Government employees

Government agency representatives should be given access to complete official visits/inspections. Government agency representatives should present their valid government organization ID card to security personnel or competent authority prior to entry.

Granting government employees' entrance to a facility does not alleviate them from following safety and security protocols. We know of one intrusive EPA inspector who decided that a particular installation was somehow fudging their emission data, and in her eagerness to inspect the incinerator site, she ignored safety and security protocols by refusing to sign in and demanding immediate and unaccompanied access to the site. As a result, she was banned from the site, and the enforcement of that ban was upheld because of her violation of security and safety protocols.

Access to the site is conditioned on observing and following appropriate security, safety, and other protocols.

Vessel personnel access through a facility

Vessel personnel (crewmembers, agents, contractors, vendors, and passengers on freight vessels) should not be permitted to depart or arrive by way of the facility unless their identification is provided in advance.

If not, entry through the facility should be prohibited until authorized by proper security personnel or competent authority in accordance with the facility security plan. All passengers and crew (on passenger ships) should be allowed to depart the vessel in accordance with INS rules. They should proceed directly to their place of work or out of the terminal.

Search requirements

All persons, packages, and vehicles entering or leaving the facility should be subject to search by security personnel or competent authority. Signs should be posted advising personnel of this requirement prior to entry. Random inspections should be conducted on at least 5% of those entering the facility while the facility is at security level I. This excludes containerized cargo.

Acceptable identification

ID cards should be a tamper-resistant and laminated photo identification card. Identification cards should show the relevant details of the holder, for example, name, description, or other pertinent data, and are to be issued by an appropriate control authority such as the Pacific Maritime Association, port authority, facility operator/ owner, labor organization, or government agency. Acceptable identification includes:

State-issued driver's license and other identification

- ID card issued by a governmental agency
- Passport
- ID card issued by facility operator/owner
- Labor organization ID card

   Exceptions: Alternatives may be accepted when worked out by facility security personnel and may include an escort if no photo ID is available and employee background checks (reserved).

Access control

Armed guard/local police department response   The facility should have armed security personnel with the authority to prevent and/or respond to unlawful acts, detain trespassers, and protect the facility, or the facility (coordinates with the port authority) should establish a working arrangement with local law enforcement that should ensure a response time of 10 minutes or less.

Gates   All perimeter gates should be locked, secured, or guarded at all times.

Deliveries   Deliveries refer to supplies and services unless otherwise noted. All packages entering or leaving the facilities should be subject to search by security personnel or competent authority. Signs should be posted advising personnel of this requirement prior to entry. This does not include cargo containers. Arrival of deliveries should be scheduled in advance. Where not scheduled in advance, deliveries should be prohibited entry onto the facility until approved by competent authority. This does not include cargo containers. The facility operator should establish procedures to ensure the validity, safety, and security of all HAZMAT shipments prior to acceptance.

VESSEL ARRIVAL AND SECURITY PROCEDURES WHILE MOORED

Facilities should not permit unscheduled tugs, barges, or other vessels to berth alongside without prior notification from the port authority or facility. Arriving vessel crews should be advised of a facility's security level. Vessel crewmembers should not be permitted to depart or arrive by way of the facility unless their identification is provided and verified. Vessel agents should schedule vendors and vessel visitors in advance. The facility should provide a means for the vessel to contact facility security.

INTERNAL SECURITY

Vehicle control

Facility management should develop vehicle access controls. Where possible, establish designated parking areas away from restricted areas. Where practicable, establish exclusionary zones to protect buildings or other potential high-value targets. Fully describe the measures implemented and standards used in the facility security plan. The following guidelines apply:

Automobiles approved for entry onto marine facilities should be controlled regarding their destination and parking.

All vehicles entering or leaving the facilities should be subject to search by security personnel or competent authority. Signs should be posted advising personnel of this requirement prior to entry.

Parking within the facility should be tightly restricted and should be authorized by a strictly enforced gate pass and/or decal system.

Passes or decals should be color or otherwise coded to further restrict access to authorized times and locations.

Parking for employees, dockworkers, and visitors should be restricted to designated areas that are fenced and outside of the cargo handling and designated storage areas.

Parking for vehicles authorized on facility grounds should be restricted largely to port authority, carrier, maintenance, and commercial and government vehicles that are essential within the facility. Parking for these vehicles should be restricted or clearly marked designated parking areas within the perimeter of the facility.

Temporary permits or passes should be issued to vendors and visitors for parking in designated controlled areas.

Rail security

Rail gates that allow access to a terminal should remain locked at all times, unless open and manned for passage of rail cars.

Key/ID/access card control

Controls should be implemented for all keys, facility employee ID cards, cipher locks, and computer systems. Key/ID/access card controls should be implemented to delineate which personnel have access to specific areas. A master ledger should be maintained that records the legitimate holder of each key copy, issuance of which should be controlled by management or security personnel. Locks, locking devices, and key control systems should be inspected regularly and malfunctioning equipment repaired or replaced. Only case-hardened locks and chains should be used, with chains permanently attached to fence posts/gates.

Computer security

Formal guidelines for computer security should be in place for each facility. Computerized information access should be password controlled and should be restricted on a need-to-know basis, which would include dissemination of information no sooner than required. Facilities should take steps to prevent facility equipment from being accessed by nonauthorized personnel.

Security rounds

Security personnel should conduct roving safety and security patrols specific to a facility's layout including the areas of waterside access.

Security personnel should conduct rounds at least once in a 4 hour period at varying times to prevent predictability. Adequate recordkeeping of the security rounds conducted should be available for inspection.

PERIMETER SECURITY AND RESTRICTED AREAS

The facility should establish restricted areas that control and channel access, improve security, and increase efficiency. This should provide degrees of security that are compatible with the facility's operational requirements. Examples include:

- Alarm/surveillance system control
- Power supply and lighting control systems
- Computer servers and storage devices

Barriers

Perimeter areas should be cleared of vegetation and debris that could be used to breach fences. Natural barriers such as water, ravines, etc., can sometimes be effectively utilized as part of the control boundary rather than fences. If used, natural barriers may require supporting safeguards (i.e., security patrols, surveillance, anti-intrusion devices, lighting) especially during high-threat period (security levels II and III).

Fencing

Compliance with IMO Circular 443 or US Customs Regulations is considered equivalent. Fence perimeters should meet the following minimum:

- Security fences and other barriers should be located and constructed so as to prevent the introduction of dangerous substances or devices. Fencing should be 8 ft high, 9 gauge galvanized steel, of 2 in. wide chain link construction topped with an additional 2 ft barbed wire outrigger consisting of three strands of 9 gauge galvanized barbed wire at a 45° outward angle above the fence.

- The effectiveness of a security fence against penetration depends to a large extent on the type of construction employed in its building.
- The bottom of the fence should be within 2 in. of the ground.
- Security fence lines should be kept clear of all obstructions.

LIGHTING

Facilities should be illuminated at least to the level of twilight and should be provided sunset to sunrise. The minimum standard for illumination should be 1 ft candle at 1 m above the ground. Dock work areas; container unloading and loading areas; waterfront, perimeter, and restricted areas; and all access points should have 5 ft candle illumination.

Lighting should conform to federal regulations (e.g., OSHA) and comply with voluntary agreements such as the US Customs Sea Carrier or Super Carrier Initiatives (if applicable). Updated lighting technology should be used, such as high-pressure sodium, mercury vapor, or metal halide lighting. Lighting should be directed downward and away from guards or offices or navigable waterways and should produce high contrast with few shadows.

SECURITY ALARMS/VIDEO SURVEILLANCE/ COMMUNICATIONS SYSTEMS

Alarms

Intrusion detection systems and alarm devices may be appropriate as a complement to guards and patrols during periods of increased threat. All control and switching systems for alarms and communications systems should be in a restricted access area. Alarms may be local, that is, at the site of the intrusion, provided at a central location or station, or a combination of both. The standard response time by facility personnel to alarms should be no more than 5 minutes.

Video surveillance

Closed-circuit television cameras can be used as a part of the facility security system. When used, cameras should be placed at main entrances and exits and in areas with high-risk and/or high-value cargo. Cameras should be able to record at relatively low levels of light and should have a remote control zoom lens capability when used for surveillance.

Cameras should have video tape recording capabilities and be capable of being monitored at the same time. Cameras should be positioned, with a recording mechanism to video record vehicles and pedestrians entering and exiting the facility.

Communications systems

Security and communications system should be tested once per shift, and a record of results maintained. A means of transmitting emergency signals by radio, direct-line facilities, or other similarly reliable means should be provided at each access point for use by the control and monitoring personnel to contact the police, security control, or an emergency operations center in the event assistance is required.

The facility should ensure adequate backup/emergency power supply in place to operate security and communications systems when the primary power is interrupted. The facility should further have dedicated emergency/security communications system in place. Each person on the security force should be issued equipment and trained on the system operation.

TRAINING AND SECURITY AWARENESS

Security force personnel (whether permanent or contracted) should complete training and qualification programs established by the facility operator and described in the facility's security plan that provides the knowledge and training to properly secure the facility.

The training program should be job specific for all personnel, both management and nonmanagement. The program should at a minimum include the following elements:

- Law enforcement and security guidelines
- Company policies including the security plan and response procedures
- Prevention, detection, and investigation of criminal activities
- Reporting of threats or actual criminal and terrorist activity
- Communications and surveillance system operation
- Procedures for notifying all facility personnel when higher security levels are imposed

Employees should be given an annual awareness training refresher to ensure that they have an up-to-date working knowledge of the facility's emergency plan including security procedures and procedures for notifying law enforcement agencies. Training program should be reviewed and personnel qualifications certified annually.

FLOATING BARRIERS

There are several relatively new floating barrier systems that can be towed into place around the vessels to prevent the approach of excluded watercraft. Generally, these barriers contain highly visible floats that have submerged cables and/or nets. Any small vessel trying to penetrate the barrier will snag on the cables and either have its bottom ripped out or the propeller destroyed. The ASTM F-12 Committee has developed a set of criteria for exclusion barriers for small watercraft up to about 40 ft in length. The Bureau of Reclamation of the US Department of the Interior has developed contract language for certification of boat barriers (see note 2). This standard requires that an 8500 pound boat, traveling at 40 knots, be stopped within 10 m of the original position of the barrier. Because vessel exclusion barriers are costly, the decision to install a vessel barrier should be based on a site-specific risk assessment.[2]

NOTES

1 Hollida R. Examining maritime risks and defenses. Appeared in http://www.maritimesecu rityoutlook.com/index.php/features/46-seaportthreats1 and http://www.maritimesecurity outlook.com/index.php/features/38-vulnerabilities.
2 According to the Department of Homeland Security, the following additional resources contain information about watercraft exclusion barriers: (i) Department of Defense Security Engineering Facilities Planning Manual, Unified Facilities Criteria (UFC) 4-02001, September 2008; (ii) US Bureau of Reclamation, Directive and Standard SLE 03-01, Standard Criteria and Procedure for Certification of Boat Barrier Systems, Appendix A (http://www.usbr.gov/recman/sle/sle03-01-AppA.pdf); and (iii) ASTM F2766-11 Standard Test Method for Boat Barriers.

CHAPTER 6

BASICS OF CYBER SECURITY

In a modern environment, there are several types of communications: wire (plant and outside telephone), fiber optic (data and voice), radio (pagers and mobile radios), and satellite upload and download links. All are susceptible to interference and attack including tapping and monitoring (may be illegal, but when has that stopped anyone with mal intent?), signal interruption (jamming, cutting lines, establishing false commands), spoofing and misdirection (think radio commands and requests for assistance or misdirection), and introduction of viruses, worms, etc. through these channels into the other hardware and software that operate both analog and digital equipment.

COMMUNICATIONS LIFE CYCLE

All communications security goes through a life cycle, and this life cycle is only a slight modification of the basic security life cycle, which has been presented before in a previous chapter. It indicates the need for continuous review and improvement (Fig. 6.1).

The process of cyber security is continuous; it must be continuously updated. As new programs and cyber intruders develop and exploit new vulnerabilities in computer programs and systems, the network defenders have a continuous struggle to defeat their attempts. High-profile attacks by China and private hackers have exposed vulnerability in even the best and most secure systems including the New York Stock Exchange, the major credit card companies, and many of the major banking and financial houses. Some data breaches are made public; most of them, however, are not unless vital information is compromised.

A recent study by Symantec's research arm Norton indicated that in 2012 there were 18 victims of cyber crime per second or a total of 1.5 million victims per day. The total value of cyber crime worldwide is $110 billion USD/year.[1] According to the report, approximately 40% of the people affected by cyber crime do not use strong passwords. Additionally, there are malware and other programs where the unsuspecting are enticed into accessing websites that implant tracking cookies or programs that log keystrokes and lock the user's computer.

Industrial Security: Managing Security in the 21st Century, First Edition. David L. Russell and Pieter C. Arlow.
© 2015 John Wiley & Sons, Inc. Published 2015 by John Wiley & Sons, Inc.

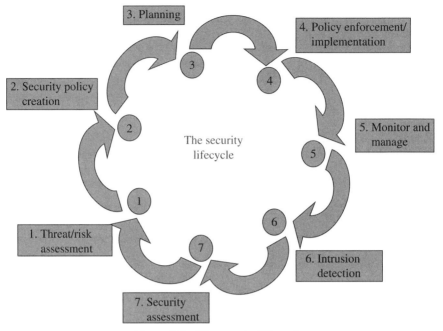

Figure 6.1 The security life cycle.

SOME SOLUTIONS TO THE PROBLEM OF CYBER CRIME

There is no *one* solution to the problem of cyber crime because of the changing nature of the problem. Increasingly, more and more of cyber crime is accessed through mobile devices by individuals who use their cellular phones on public networks. But, businesses are equally vulnerable, and networks are getting harder to defend as they become more complex.

The following general recommendations should be incorporated into any company in an effort to provide basic cyber security. These are some of the basic elements, and the list is far from all-inclusive, but it does address the basic elements.

General recommendations

- Have a written policy on communications about the purpose of the communications and the types of communications that will and will not be permitted on company machines.[2]

- Analyze and plan the types of communications your company wants and needs. This should include such elements a spectrum use and types of CCTV and radios used within a plant.

- Study and analyze the appropriate equipment for your system—it should permit implementation of the written policies.

- Enforce the policy. Internet tools including key logging, packet sniffing, and similar programs are legitimate ways of finding out what employees are doing including reading e-mails. *However, make it clear to the employees as they are hired what the policy is, what will be tolerated, and what will happen if violations are encountered.*

- Once the equipment is in place, then it must be operated, managed, and maintained. Nothing is as frustrating as an unreliable company-wide Internet or intranet system.

- Back up the system regularly and also remotely.

- Keep personnel and sensitive files on separate dedicated restricted computers. Employee information and salary data and associate materials must be safeguarded. In the event of a system penetration, the data should be secure.

- Employ a security encryption system for sensitive and business-related data. Even relatively simple encryption systems can make stolen data worthless.

- Radio communications and Wi-Fi and all broadcast information should be encrypted. This should include telemetry signals from sensors and other process control sensors.

- Change passwords regularly. Most people do not remember passwords from 1 day to the next, even when they create them. There are electronic methods that will enable employees to keep sensitive passwords on read-only flash drives.

- Decide which computers should have hardware-permitting input by CD, DVD, or flash drive. Operator consoles and other sensitive areas should not have removable drives nor flash drives where software can be added or removed.

- Central operating systems and networks are, in general, a good thing because they promote efficiency—but they may also represent a potential for security breaches, especially if someone can access critical functions and programs from outside the company.

- Critical operations and processes used in plant control should have supervisory access only from specific computers.

- Operator consoles should not have Internet access. Operators with Internet access can be distracted from process monitoring during their work shift if the Internet is available.

- If the system is designed properly and tested periodically, it should be able to detect intrusions—but that requires a willingness to test the security of the system.

- Every site should have a security assessment team that includes inside and outside professionals.

- The security assessment should also include a threat/risk assessment that reboots the assessment process again when there is something that causes new or different threats and risks.

COMMUNICATIONS SECURITY

Some of the first philosophical questions that must be answered when looking at a secure communications system are where the communications are taking place, who is generating information, what information they are generating, and who needs to receive the information.

The purpose of a communications system is to facilitate an interchange of ideas, which may be mobile or not, and collect and distribute them collectively to where they are to be acted upon. Some examples include:

- The shipping department must be able to connect to the receiving department for routing vehicles. Both must be able to communicate with the guard station and warehouse to authorize the receipt and storage of a shipment.

- The telephone system provides a central nexus for all communications in the plant (Centrex, Switchboard, or something else).

- The guards must be able to summon assistance when required much like the fire brigade and the crash cart and ambulance.

COMMUNICATIONS AS TRANSACTIONS

When we start to look at the communications as transactions, they take on a different view. In that manner, we can prioritize their importance and relate the facilities to their purpose. Communications transaction security requires both the sender and the recipient to be secure. Are the following secure and practical and why?

- Guard making rounds communicates to central dispatch in front of the plant.
- Guard calling for help on the telephone.
- Laboratory calling production operations with results of tests.
- Central communications station next to the boilers in the plant use the plant stack as antennas for plant-wide communications.
- Workers are summoned by pagers.
- Workers and guards informed of problems by plant radio from central station.
- General manager phones a worker in another building to come to a meeting.
- Lila (purchasing) phones Mary in shipping.

TELEPHONE SYSTEM SECURITY

Some or all of the following questions should be asked and answered if one is to have a secure telephone system:

- Are the communications centralized through one vulnerable area such as a telephone trunk line and PBX switchboard, or are they distributed?
- Where is the central trunk line located with respect to the plant?

- Is the central communications line accessible outside the plant? Is the phone cable buried and is the switch box in a hidden location?
- How easily can someone with mal intent gain access to the phones, and how easily can they be disrupted or monitored?
- If a telephone truck showed up at the front gate or parked at the central switch-board or junction box, unannounced, would anyone notice?
- If the local phone company had valid credentials, would your guard force allow them entry into the plant without checking with the telephone company dispatch to verify that the persons were legitimate and without being aware of a communications problem where outside help had been called in?

RADIO COMMUNICATIONS

Every large plant uses a plant radio, some use cellular telephones, and others still use pagers. Analog systems are easier to break into and spoof than digital systems, but there are still a lot of analog systems in use. The following should be tested:

- Is the central radio secure or vulnerable to attack and being disabled from within or without the plant fence?
- Can anyone with a radio of the right frequency gain access to (monitor or interrupt) plant communications?
- Are plant public broadcasts (all radio) transmissions encrypted or in plain language?
- How easy would it be to steal or duplicate a plant radio?
- If you have a centralized dispatch system, do you have a backup location in case of an emergency? Does that alternative station have the same capabilities that the regular station has? In other words, is it a full duplicate station and does it have full capabilities? Is it used and maintained periodically just to make sure it is functioning?
- Would a power failure cripple your ability to communicate either with inside or outside facilities?
- Would an attack that destroys the plant radio house eliminate your ability to obtain plant-wide assistance or summon or control the workforce?
- Is there an alternative method of summoning outside assistance if the main trunk telephone line is cut?
- What happens if the local cellular telephone tower is destroyed? Or what happens if someone attempts to use cell phone jammers?

DIGITAL COMMUNICATIONS

Many plants have common systems that are wide open to attack from the Internet and other sources. Just because the transaction involved in digital exchange is usually incomprehensible to our ears and eyes does not mean that the digital transactions are unimportant. Transactions between machines are much like human transactions, in that there are a call and response to initiate the transaction and code checking when the recipient recognizes that the transaction is directed toward it. Then, there are an interchange of data and a closure agreement when both parties sign off.

Sometimes, the conversation is dedicated (telephone and SCADA), other conversations are continuous and do not sign off, and some digital conversations are shared, while others are not. It all depends upon how the data are transmitted, the devices, and the encoding. We often do not pay attention to it, because it is automatic. An example is the old dial-up modem. When it connected with the service provider, it provided a code of dial tones, and then when the provider responded, the system would produce something that sounded like static but that is digital communication.

When we examine the challenges of providing adequate cyber security and the integration of World Wide Web pages and websites into a server, even by e-mail rerouting, it is easy to see how attacks can take place via the Internet and through digital and analog communications via Wi-Fi and wired communications. The variety of digital attacks alone are virtually endless. A recently established website put up by T-Mobile (Deutsche Telekom) lists the type of attacks occurring each second of the day. A recent study by SYMANTEC (INTERNET SECURITY THREAT REPORT -2014 http://www.symantec.com/content/en/us/enterprise/other_resources/b-istr_main_report_v19_21291018.en-us.pdf) indicates that businesses have a 1:3 to a 1:5 chance of being cyber attacked, (and that is based on the attacks that were caught and foiled). A similar report by Shane Shutte in September 2014 (http://realbusiness.co.uk/article/27859-there-are-now-117339-cyber-attacks-per-day) indicated that there are about 1.4 cyber attacks per second, but the article is unclear as to whether that was just for the United Kingdom or worldwide. Other statistics have indicated that the worldwide number of attacks is between 3 and 4 attacks per second. It is important for any company to train their employees to avoid the various traps and malicious e-mail scams and web-based programs that are prevalent on the Internet. Every company needs a very good firewall and IT department to keep their Internet functioning.

CYBER SECURITY

Vulnerability assessment

The basic principles for cyber security are the same as with physical and general security. It starts with a vulnerability assessment, but we also need to recognize the transactional nature of the communications. The elements of that vulnerability assessment start with the following:

- Risk assessment and threat identification
- Assessing probability of attack
- Identifying and ranking the criticality of the communications transactions
- Identifying the most critical and the associated costs if they are interrupted or lost

- Identifying the effective actions required for their prevention and the associated costs
- Evaluating the costs of prevention against the benefits obtained
- Documenting the findings and preparing an action plan
- Implementing the action plan and monitoring the results.

Unknowns and alternatives

There are a number of challenges to this approach including a significant lack of current and in-house data (we do not know what we do not know!)—we lack information about the criticality and nature of both the information and the nature of the communications until we complete the assessment and do not know which controls or communications or cyber controls are critical.

We also lack information about effective alternatives and have to realize that uncertainty and lack of information are a constant in that the Internet and computer systems are both growing and dynamic, that our efforts represent a snapshot in time, and that we probably will have to repeat the assessment in response to changing operating systems and changing hardware.

The difficulty with the development of alternatives lies in the costing precision of the alternatives. In industry, cost is not the only factor but is an important factor. Evaluation of alternatives requires some guesswork, and the greater the focus on development of a precise alternative scenario, the more expensive it becomes to develop that scenario. Preparation of accurate cost estimates for evaluation of alternative scenarios is often very difficult, time-consuming, and expensive because it requires evaluation of a number of factors.

Among those factors are process evaluation and costing, lost alternatives costs, and human and machine effectiveness (efficiency) calculations (there is no given standard).

HOW TO PERFORM THE VULNERABILITY ASSESSMENT

There are any number of ways a vulnerability assessment can fail, most notably when the credentials of the assessment team are questioned or when senior management says, "I don't like it," "This can't be right!," or "I don't believe it!"

There are a number of things that an assessment team must do in order to have a chance at developing an assessment that will be considered and utilized. These critical success factors include the following:

Critical success factors

- Obtain senior management support and involvement.
- Define the scope of the assessment.
- Approve action plans after they are developed.

- Designate focal points and task force.
- Preferably one individual who is familiar with company policies and procedures and the "way things are done."
- Individual serves as coordinator for centralizing information and directing resources.
- Weekly group meetings.
- Involve those responsible for collection and management and use of data.
- Designated focal points enhance the quality of the development and efficiency of the risk assessment.
- Tools and programs should be evaluated by those who will use them.
- Development techniques can be cross applied to other areas.
- Insure that language and meanings of terms are consistent throughout the development process and across the plant.
- Insure that reports are delivered on time and in standardized formats.
- Insure that expectations of senior executives are met.
- Involve specialists:

  - Involve specialist disciplines such as engineering, business, and operators as well as IT people.
  - Business managers often have the best understanding of the criticality and sensitivity of business operations and of data systems supporting them.
  - Technical and security people often understand existing system designs and vulnerabilities.
  - Include auditors and financial people.
  - Equipment vendors probably should be excluded at these meetings because they have specialized agenda.
  - Where possible, conduct risk assessments with in-house personnel.
  - Team approach used to conduct external risk evaluation from federal agencies and CERT (Carnegie Mellon Computer Emergency Response Team) (www.cert.org).
  - Hold business units responsible for initiating and conducting their own risk assessments.
  - Limit the scope of individual assessments.
  - Limit the scope to individual business units rather than a large all-encompassing scope.
  - Segment operations into logical units.
  - Different operations have different risk levels.
  - Identify shared risks with associated infrastructure—that is, e-mail, common files, common programs, etc.
  - Document and maintain results.

- ◦ Force accountability for managers.
- ◦ Make auditing performance easier.
- ◦ Provide a starting point for subsequent assessments.

Optimum assessment team size

This stated, it is probably best to have an assessment team that is between 5 and 10 persons. Too few and the job becomes enormous; too many and the group becomes unwieldy to manage.

The vulnerability assessment process will report and develop procedures to help reduce or eliminate cyber vulnerabilities. Because this is often part of corporate "policy and procedures" (P&P) and will wind up in the P&P book, it must be a formal written set of statements that clearly outline the steps to take in response to real vulnerabilities or situations.

COMMUNICATIONS PROCEDURE DESIGN: HINTS AND HELPS

- • A written procedure must be clear, in plain language, and easily understood, even if that means redundant with respect to some actions.
- • Do not let your legal team anywhere near the procedures because they will screw it up by applying legal principles and arcane language.
- • Procedures, especially those that involve the entire company, must be consistent across diverse business units.
- • Must be developed in concert with others rather than in isolation (no need for "reinventing the wheel").
- • Keep in mind that the sharing of information, especially about real or perceived vulnerabilities and attacks, is important.
- • Identify parties responsible for initiating and conducting risk assessments.
- • Determine who has to participate.
- • Get agreement on steps to be taken.
- • Determine in advance, if possible, how disagreements are to be resolved.
- • Identify which approvals are required.
- • Determine how the assessment is to be documented.
- • Determine how documentation is to be maintained.
- • Determine who gets the reports.
- • Determine who can authorize recommendations.
- • Standardize reporting formats, that is, tables, lists, questionnaires, and standard reports: KISS, or keep it simple, stupid!

BENEFITS: IDENTIFIED

The identified benefits from providing a cyber security risk assessment and developing policies and procedures to address those risks include identifying the risks on a continuing basis; helping employees and personnel understand the business; helping employees avoid risky behaviors and practices; alerting employees to be aware of suspicious events and practices coming in via e-mail, Internet, and other forms of digital communications; and providing an effective way to communicate risk information to specific business units.

Example

The following is a risk assessment process diagram developed for a chemical company (Fig. 6.2).

One large chemical company has a risk cyber risk assessment policy in line with their general risk assessment policy. The oil company reported that they initiated a risk assessment every 3 years or after any significant loss or incident. In either case, they notify their regional corporate security coordinator/manager (RCSM) to start the process. The regional security coordinator initiates the process with the corporate security who reviews the budget and project documentation. The RSCM develops the risk assessment team, involves the business unit managers and other executives, and prepares a risk assessment execution plan.

A risk assessment team composed of five to eight persons is formed, and they select a leader from a different business unit within the company to insure objectivity.

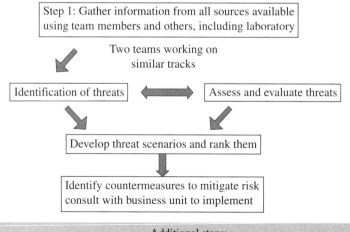

Figure 6.2 Risk assessment team assignments for a chemical company.

The assessment team conducts interviews and develops a risk assessment plan. The plan is iterative in nature and the final plan must receive approval from the business unit manager.

Meanwhile, an independent group at the corporate level (sponsored by corporate security) develops additional risk information and feeds it to the team. The team develops a risk assessment baseline that includes external threats, internal threats, system-induced threats, and other threats.

The risk assessment team effort should be between 1 and 2 weeks but less than a month. Between 60 and 75% of the team's efforts are directed at collection of the data; the balance is for the preparation of the report. The team will conduct 30–50 interviews, with each interview being at least 1 hour to explore the issues. Written submissions and examples are encouraged as long as the information is not too volumetric to handle. Team members may not interview coworkers to maintain objectivity.

The last phase of the team effort is to develop at least 10 or more problem scenarios and must consider the ways in which current procedures and applications may compromise security and resources and damage the company. The team must also consider the possibility of disclosure of confidential information, loss of data, inability to communicate with outside and internal resources, and operations. Developed scenarios may include employee motivations for disclosure of proprietary information or secrets for personal gain or for relief of financial stress. The written report is then tabulated in a threat matrix as shown in the following.

CYBER THREAT MATRIX: CATEGORIES OF LOSS AND FREQUENCY

Categories of loss I through IV:

- Death, loss of critical information, system disruption, or severe environmental or other damages
- Severe injury, loss of proprietary information, occupational illness, and major system or environmental damage
- Minor injury, minor occupational illness, and minor damage
- Less than minor injury, occupational illness, or less than minor system or environmental damage

Categories of frequencies A–E:

- Frequent, possibility of frequent incidents
- Probable, possible isolated incidents
- Occasional, possibility of occurrence sometime
- Remote, unlikely
- Improbable, virtually impossible

The threat matrix looks like this (Fig. 6.3):

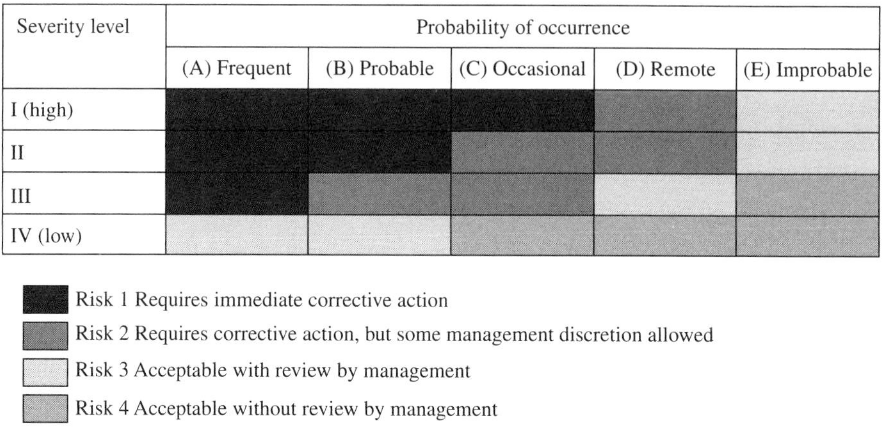

Severity level	Probability of occurrence				
	(A) Frequent	(B) Probable	(C) Occasional	(D) Remote	(E) Improbable
I (high)					
II					
III					
IV (low)					

■ Risk 1 Requires immediate corrective action

■ Risk 2 Requires corrective action, but some management discretion allowed

□ Risk 3 Acceptable with review by management

▨ Risk 4 Acceptable without review by management

Figure 6.3 Threat matrix for cyber security occurrences.

For each occurrence, a ranking and a rating are developed, and alternatively, internal software is used to help develop corrective actions based on a list of security controls and provides related cost estimates. The team should develop a list of possible corrective actions.

For each scenario requiring risk reduction, the team should identify one or more corrective actions from the list of alternatives they have developed and then select the most effective correction action based on the effectiveness of possible control in reducing probability or severity of incident and cost.

Finally, the team should develop corrective actions and prepare an exit briefing and a draft report that each team leader reviews. When all the changes are finalized, the report is implemented and the changes are submitted to middle management for their consideration and implementation. Middle management will receive the report and implement the changes. Corporate security will monitor the progress of the implementation changes.

SETTING UP INTERNET SECURITY

Internet vulnerability is potentially one of the greatest threats to cyber security. Through a combination of e-mail, web browsers, and other programs, there are substantial security holes in any web-based system. They include (i) Trojan horses, (ii) worms, (iii) malware, (iv) denial of service, (v) attacks, (vi) keystroke loggers, and whatever new that is under development at any given moment. According to various sources on the Internet, even JPEG files and other types of graphic elements can contain hidden controls that can corrupt or take over a computer's operating system.

Part of the problem is that the Internet Explorer is particularly vulnerable precisely because it is integral to the Windows operating system and is impractical to remove or disable the system. Apple systems are harder to attack because the registers in the operating system are individual to the machines, whereas Windows registry files are all similar.

External versus internal testing

Many computer security professionals recommend starting with an external Internet assessment for the purpose of checking the vulnerabilities of the computer system and individual computers. The external assessment is also known as a "perimeter test," and it is conducted from outside the network. This emulates hacker attacks, seeking ways in which the system can be penetrated.

A second type of test is the internal test, "and emulates the threat experienced from internal staff, consultants, disgruntled employees, or, in the event of unauthorized physical access or a compromise of the perimeter security. These internal threats comprise more than 60% of the total threat portfolio.[3]" An Internet assessment will not address every threat to your network, but may catch most of them. Threats from remote access servers and connections to third parties are generally not detected by the assessment.

Security focus

Starting with the customer's own website(s), they are mined for information about the customer. The primary objective is to derive the DNS domain names that the target uses and map them to the IP addresses to be investigated. DNS domain names often come from the e-mail address or the company's name.

Using search engines, search all instances of the company's name. This provides links to the company's own site (from which DNS domain information can be easily derived) and provides information about mergers and acquisitions, partnerships, and company structures.

Using a tool like HTTrack, dump all the relevant websites to disk. Then scan those files to extract all mails and HTTP links, and parse them to extract more DNS domains.

Browser and domain security

Use the various domain registries. Tools like geektools.com, register.com, and others can often be used:

- To verify whether the domains we have identified actually belong to the organization being assessed.
- To extract any additional information that may be recorded in a specific domain's record.
- For example, you will often find that the technical contact for a given domain has provided an e-mail address at a different domain. The second domain then automatically falls under the spotlight as a potential part of the assessment.

Some of the registries provide for wildcard searches. Such a search can help to identify all the domains that may be associated with the company ABC Apples Inc., for example. The object and output of this work is a comprehensive list of DNS domains that are relevant to the target company. It may require several searches and trials and revisions to get a complete list.

One of the first things that an attacker is looking for is the IP/name mapping for the target company. Name server and mail exchange records often contain this information. Because the IP addresses tend to be grouped together, it is often easy to find the range of addresses for a specific network. This makes it easy for an intruder to conduct a "ping" scan to find out which specific IP addresses are active.

Almost every machine on the Internet works with a series of little Internet locations called ports. Ports are used to receive incoming traffic for a specific service or application. Each port on a machine has a number, and there are 65,536 possible port numbers. An IP address can be probed using a "port scanner," a freely available software utility that tries to establish a connection to every port in a specified range.

If a network is on the Internet, there are a number of possible active ports through which an attack can occur. Assuming that an IP address is active on the Internet, for a reason, the most likely port addresses are mail servers on port 25, web servers on ports 80 or 443, a Microsoft client on port 139, remote mail server on port 587, and a DNS server on port 53.

An intruder or hacker can access a network in any number of ways. Configuration errors and mismatches between various software components may provide an entry to the system. In some instances, a "stack overflow" occurs when a program or subprogram uses excessive amounts of memory beyond the allocated amount. This type of error can be accidental or deliberate and may be caused by an infinite loop program that is very easy to write. Because programs are complex, a minor transfer error can cause an opening that is exploitable. Also, programmers have been known to leave "backdoors" in their programs for the purpose of service, correction, and access without having to go through the program authentication.

http://www.SecurityFocus.com and http://www.SecuriTeam.com contain information on security vulnerabilities. An Internet search engine will also be able to return operating system vulnerabilities, allowing one to exploit those areas if they have not been blocked or disabled. There are also vulnerability scanners that check for installation of best practices on a system and best configuration for desired results. Companies like ESET, Symantec, McAfee, and BindView all make vulnerability scanners although they have different names. There are even some open-source scanners that are freeware, such as Nessus from Renaud Deraison. Currently, a search engine listing for network vulnerability scanners returned with 730,000 hits or potential sources to investigate.

Data encryption

The issue of data security can be a complex one. Data encryption is often time-consuming and carries an overhead, but it can prevent an outsider from seeing critical data. That said, it is not necessary to encrypt everything, and the network and communications analysis should indicate the types of communications that should be encrypted and those that should not. For example, a highly sensitive e-mail between coworkers might be encrypted, especially if it or the attachments contain business-sensitive information. This requires a "two-key" type of security system where only the person receiving the message can decrypt it.[4] Management of a secure communications system while maintaining a network with wide access is often a challenge. Most security professionals will recommend a 128 bit key encryption system.

A part of the company's security and intranet system should also address encryption for any devices that operate or that can be controlled over a network. It must support both sensor and control systems and node-to-node communications directly or through intermediaries. In industrial control systems, all of the communications models of common field I/O programming must be supportable, and the system must insure that the failure or compromising of a single node will not bring down the other nodes or the system.

Authentication is also a significant part of the system. Generally, a dual-band system for control systems where the authentication is transmitted by a different mode such as radio or cable or infrared systems generally insures control system security. The importance of this issue cannot be stressed strongly enough. In order to be secure, the data transmission rate between sensors and control points should be high and should be able detect loss of information packets and loss of synchronization.

CYBER SECURITY TOOLS

The US Department of Homeland Security has developed the following tools for use. Some may be export controlled. The National Cyber Security Division's Control Systems Security Program (CSSP) (http://www.sans.org/press/dhs-inl-win-ncia.php) is designed to reduce cyber risks. They have developed the following tools:

- Catalog of Control Systems Security: Recommendations for Standards Development.
  Control System Cyber Security Self-Assessment Tool (CS2SAT).
- CSSP documents in conjunction with the Industrial Control Systems Cyber Emergency Response Team (funded by the US Department of Homeland Security) (ics-cert.us-cert.gov/csstandards.html#control). Cyber Security Procurement Language for Control Systems.
- The ISA Automation Standards Compliance Institute is a licensed distributor of the CS2SAT program. The program will assist SCADA and process control system users in improving their security.

NOTES

1 http://now-static.norton.com/now/en/pu/images/Promotions/2012/cybercrimeReport/2012_ Norton_Cybercrime_Report_Master_FINAL_050912.pdf. Accessed 2014 Oct. 14.
2 According to Internet statistics, about 4.2% of Internet browsing is in response to individual's desires for some type of pornography. The sites that carry not safe for work (NSFW) materials should be blocked. This was documented by a 2011 study in *Forbes* magazine: http://www.forbes.com/sites/julieruvolo/2011/09/07/how-much-of-the-internet-is-actually-for-porn/. Accessed 2014 Oct. 14.

3 http://www.symantec.com/connect/articles/assessing-internet-security-risk-part-1-what-risk-assessment. Accessed 2014 Oct. 14.

4 One of the problems is deliberate retransmission. A number of years ago, before e-mail made communications easier, one of Occidental Petroleum's executives wrote a "sensitive information" memo to another, clearly marked with "DO NOT COPY OR DUPLICATE." In response to a subpoena on the subject, a file search came up with over 20 copies of the sensitive memo, sometimes in remote places in another business group!

CHAPTER 7

SCENARIO PLANNING AND ANALYSES

INTRODUCTION

In Chapter 2, we briefly introduced the concept of fault tree analysis (FTA) and network analysis for evaluation of risks and consequences. The purpose of this chapter is to dig down further into those subjects and other subjects to demonstrate practical analytical techniques for scenario planning: how to analyze a network and how to perform security and hazard analysis by various methods. In this section, we will analyze the formulation and consequence of various types of incidents and demonstrate some of the practical ways to eliminate or reduce their consequences. We will look at fault tree and network analysis in a bit more detail, then move on to the requirements of the US Department of Homeland Security (DHS), and finally consider consequence analyses and discuss the methods of evaluation and some of the software available for that analyses.

In this chapter, we will go further into specific methods to discuss risk assessment and hazard analysis techniques and disaster prevention as it applies to process operations and plant security and security management. It should be noted that in many instances, there is a substantial overlap between physical security and plant safety. No, we are not talking about worker safety, but process safety. The safety of the process has inherent implications that can affect the plant as a whole, and in the event of a disaster or serious process-related accident, the plant security force will play a major role in the response and in providing security to the plant and the personnel during and after the incident.

The analytical techniques between plant safety and plant security are often identical, and the principal distinction is that plant security is often outward directed, whereas plant safety is more internally focused on the plant operations. However, the distinction may become moot when there is either a serious plant accident, a sabotage, or a damaging attack from outside the plant.

When we get to the later portions of this chapter, those dealing with contingency planning and emergency response, we will discuss the outlines for emergency response *from the perspective of the security force*. The types and kinds of response are dictated by the incident, and the response must be coordinated by trained individuals who have the plant, its people, and the community uppermost in mind.

Industrial Security: Managing Security in the 21st Century, First Edition. David L. Russell and Pieter C. Arlow.
© 2015 John Wiley & Sons, Inc. Published 2015 by John Wiley & Sons, Inc.

Before we start the discussion of specific analytical methods, a preliminary comment is offered. Fault tree, failure modes and effects, Bayesian network analysis, and some of the HAZOPS techniques are all related. Saying one is substantially different than another is somewhat misleading as most have a similar purpose and end result. The differences are in how the result is achieved. Most of the techniques use a similar analytical form, and each analyzes it slightly differently. The amount of work involved in development of a good analysis is directly dependent upon the effort involved, and the assumptions used in its development. The one fundamental thing to remember about the analysis is that while it looks like solid stuff, it is only as good as the assumptions that go into it. No amount of risk analysis will prevent incidents or occurrences of bad outcomes if the plant management does not follow through on the implementation of the recommendations of the risk management plan.

In this regard, one has to look no further than three of the largest industrial accidents in recent times: Chernobyl, BP Texas City Refinery Explosion, and Bhopal, India. Each of these is well documented in the chemical and risk management literature and has been extensively studied. The commonality of each of the disasters is related to multiple failures and a management disregard for the potential for those disasters. Disasters seldom come from a single event but are the result of management failures and multiple sequential violations of common sense and safety procedures.

In Bhopal, maintenance and safety procedure failures that led to a critical blind flange not being installed, coupled with decommissioning and shutdown of a methyl isocyanate cooling unit, and the temporary shutdown of the vent gas scrubber were contributing causes that led to the release that reportedly killed up to 7000 people in Bhopal, India.

In Chernobyl, Ukraine, the decision by a couple of electrical engineers, <u>not nuclear engineers,</u> to run a reactor spin-down test with the deliberate overriding of six layers of warning alarms, coupled with the lack of senior management to supervise the test, led to the Chernobyl disaster. According to an account by Rushworth Kidder, who investigated the accident, the test could have been conducted on a nonnuclear reactor with greatly reduced consequences.[1]

The third incident was the loss of life due to the explosion and fire at the BP Texas City Refinery. The US Chemical Safety Board found a number of safety and process violations that were contributory to the incident, but key among them was the lax safety culture at the refinery.[2]

FTA, MARKOV CHAINS, AND MONTE CARLO METHODS

Earlier, we discussed and briefly outlined the preparation of a fault tree analysis. Implicit in the preparation of an FTA is the assumption that we can quantify the probabilities for success or failure for each of the branches or events. Often, we cannot judge the probabilities for such successes or failures, and we have to look at other methodologies that would enable the likelihood of the larger events. This is one way of evaluating a "worst-case scenario."

Fuzzy fault trees

In Chapter 2, we described an approach to fault tree logic that used an assessed probability of an event taking place. Strict interpretation of fault trees does not assign that probability of an event as in analysis; it is either pass or fail for a fault.[3] The method illustrated used probabilities to provide an assessment of the likelihood of an occurrence. A further adaptation of that method uses fuzzy logic that can also be applied to FTA, where instead of a fixed probability being assigned (in strict FTA, the probability is either 100% true or false), a range of probabilities can be applied to each element. For example, if a particular event has between a 20 and 60% likelihood of occurrence (or any other percentage between 0 and 100%), the event could be considered probable, and the fault would be conducted to the next layer of the system for further analysis. This analysis allows us to better estimate risk in an uncertain environment.[4] The use of fuzzy logic is an improvement over the conventional "on-or-off" logic implicit in an FTA, and it uses the same analysis techniques described in Chapter 2, with the exception that it involves a lot more computation, because for each event multiple calculations are required as one steps through the calculations.

For example, if the probability for one event is between 0.2 and 0.6 (20 and 60% likely), one might have to repeat the entire set of calculations for the entire fault tree at stepped intervals—say 0.2, 0.22, 0.24,...,0.58, 0.58, and 0.60. The process would be repeated with appropriate variables on the next element. This is where computers are extremely useful in reducing repetitive and tedious calculations.

Markov chains and Bayesian analysis

Other effective ways of evaluating the likelihood of probabilities for success or failure of an attack, especially where we do not have data, are to perform a Bayesian network analysis or a Markov chain analysis. These types of analyses can be constructed considering upon the failure or reliability of various subcomponents of the plant. In these types of analysis, it is often useful to diagram a risk tree, with the probabilities for each option. In a Markov chain, each state is treated as a variable, and matrix methods are employed to describe the risk.[5] The computations on a Markov chain can become quite complex and large, and computer methods are used to develop a risk calculation.[6] *However, if a category or type of risk is not addressed, it is unknown.*

The Bayesian network analysis is one of the most powerful analytical tools, as it determines not only direct probabilities but also examines and projects the posterior probabilities based upon the existing data. The changes in any one subcomponent of the plant can then be examined to determine the impact, likelihood of success, or failure on other components of the plan data and model. This technique is extremely powerful but requires a working knowledge of statistics, and it is computationally intensive as well.

The technique may be too sophisticated for many companies, because the most difficult task in a Bayesian analysis is the establishment of the initial condition tables to make sure that they reflect the appropriate values and relationships. The network is then examined to insure that there are no conditional faults and the data are fed into a computer to produce the probabilities. This method is computationally similar to the Markov chain analysis.

There are several very good Bayesian network programs and texts[7] as cited in the section "Notes". The assumptions and the setup of the risk decision tables are key to the analytical technique. AgenaRisk is a good program that is free with the purchase of the textbook *Risk Assessment and Decision Analysis with Bayesian Networks*, by Fenton and Neil (CRC Press). Other programs include Netica (a few hundred dollars from www.norsys.com); GeNIe & Smile is freeware from the University of Pittsburgh (www.genie.sis.pitt.edu), and SamIam (freeware from http://reasoning.cs.ucla.edu/samiam/) is also worth considering.

OTHER COMPLIMENTARY TECHNIQUES

In order to evaluate the likelihood of success or failure of individual security compo-nents, and to set up a preliminary analysis of failure modes, it is often wise to consult the American Society for Quality website, www.asq.org. They have free programs that enable the evaluation of failure modes of component devices, and these failure mode analyses can be applied directly to security elements and do much of the initial work in setting up an analysis. This type of analysis is called failure modes and effects analysis, or FMEA for short.

Fishbone (Ishikawa) diagrams

Two of the easiest to use and understand FMEAs are the *fishbone diagram (ISHIKAWA diagram)* and the *Pareto chart*. The fishbone diagram should be applied where you are attempting to identify possible causes for a problem or possible methods of attack or interference. The fishbone diagram tends to help avoid falling into ruts by the organization of the diagram. A sample fishbone diagram is shown below—prepared using the free software and guidance from http://www.fishbonediagram.org/template. The website also has a fishbone diagram template for PowerPoint slides. Both are free for downloading. We have followed the basic suggestions for fishbone diagrams and have considered the elements of man power, methods, machines, metrics, mate-rials, and minutes (time) in the diagram below.

In preparing a fishbone diagram, it is important to use a team approach and group and identify possible causes. When possible causes have been identified, group them under representative headings, and then list the individual cases on each group. No idea is too unimportant to consider, and when the diagram is completed and the causes are identified, then review the diagram and highlight the most important causes.

Example The example above is for security failure at a chemical plant that caused a successful attack on a motor control center and an operation control station (Fig. 7.1). The attackers came into the plant through a remote corner of the fence near the tank farm. They were able to be unobserved for what is estimated as an hour. In that time, they managed to enter an unlocked door in a motor control room and dump nails and sand into motor vents and into oil receptacles causing a number of large motors to burn out. They also compromised an unattended operator's con-trol station, shutting down some safety interlocks and turning on pumps that they

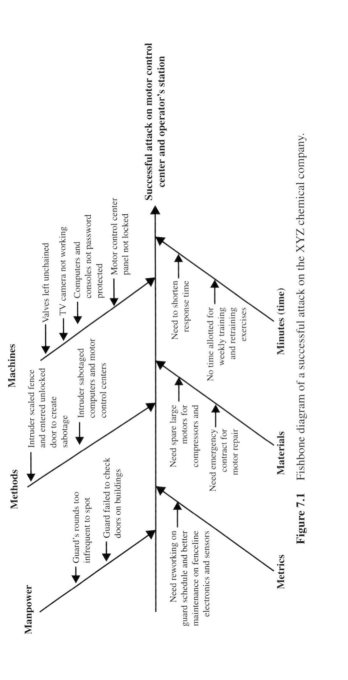

Figure 7.1 Fishbone diagram of a successful attack on the XYZ chemical company.

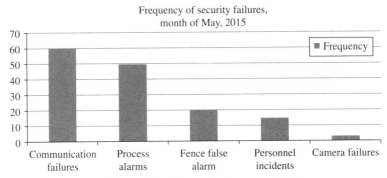

Figure 7.2 Pareto chart on security failures.

had sabotaged. The result was several tens of thousands of dollars in direct repair, plus many times that in lost production.

In the analysis, manpower, metrics, methods, machines, time (minutes), and materials are categories for failures of the security system.

In each case, the question of "Why does this happen?" must be answered. Subcauses, if any, are shown as branches of the principal element categories.

A different type of analysis for a plant attack might include the basic question of "How is the plant attacked?" The principal ribs of the fishbone could be environment, weapons, communications, cyber, transportation, storage, raw materials, supervisory controls, laboratory, and pipelines. Subcategories might include such items under raw materials: bomb in shipment, deliberate contamination of feedstock, receiving accidents, storage accidents, etc. The point is to be as thorough as possible and examine each possible method or means until the chart is complete. In some instances, more than one chart may be required for complete analyses.[8] The ASQ website also has a free download Excel© spreadsheet with the detailed Ishikawa or fishbone diagram.

Pareto charts

A Pareto chart is a different type of analysis and may be useful in analysis of the number and type of security faults occurring at a facility. If, for example, a number of security and/or safety violation incidents occurred, one would have to categorize them with respect to the type. Each type has to have a common measure over a definite period of time, such as frequency. Categories may be something like fence false alarms, sensor failures, camera failures, personnel incidents, etc. The results are converted to a bar chart with the item with the greatest frequency on the left, and the others arranged in decreasing order. A sample Pareto bar chart is shown in Figure 7.2.

SAMPLE OF INITIAL ANALYSIS

The following tables illustrate one of the types of setup for a fault tree or Markov chain or Bayesian network analysis for a plant shutdown for a major refinery or chemical plant. It is a starting point, not the finished example because it shows categories of major systems that could fail and cause a plant shutdown. The top level might be general screening categories that could cause a plant-wide failure or emergency leading to shutdown. In reality, most plants are significantly more complex than the example indicates, but it serves to list a few of the categories for further consideration and analysis (Table 7.1).

The table is designed to catch the major issues and is therefore very general. There are many other specific situations that can create a plant shutdown and a number of subcategories below each of those listed that can create a situation leading to plant shutdown. The table does not seek to address overlapping areas but is a first categorization that identifies some of the problems that can create a shutdown.

In any of the columns, the list can be further subdivided to give greater specificity with regard to the issues listed. As an example, two of the columns from the table above are further categorized in Table 7.2.

Under the heading Supply Problems, there are a number of categories that come to mind that would require further analysis:

Supply problems

Pipeline problems

Pipeline leak (PL)

Pipeline break (PB)

Pipeline corrosion (PC)

Pumping station failure (PSF)

Pumping station leak (PSL)

Incoming storage tanks

Tank overfill/spill (TOs)

Tank collapse (TC)

Tank rupture (TR)

Tank valve failure (TVF)

Tank roof problems (TRp)

Tank major spill (TMS)

Tank leaks (TL)

Tank fire (TF)

Tank capacity problems (TCp)

Finished product storage tanks

Lack of capacity (FPSLk)

Tank overfill/spill (FPSOf/S)

Tank rupture (FPSTR)

Tank collapse (FPSTC)

Contamination problems (FPSCon)

Valving problems (FPV)

Leaks (FPL)

TABLE 7.1 Plant shutdown risk analysis table of likely causes

Plant shutdown

Supply/storage problems	Electrical problems	Control problems	Process problems	Exterior problems
Pipeline	Primary electrical power	Computer systems	Thermal (boiler/steam)	Security—serious perimeter breach
Incoming storage tanks	Secondary electrical power	Controller failures	Fire	Sabotage
Finished product storage tanks	Backup electrical power	Control room problems	Explosion	Bomb
Pipeline	Motor control center failures	Faulty sensors	Pipe rupture (major)	Explosions
Port/shipping problems	Major motor driver failures (localized)	Maintenance issues/poor maintenance	Valve failures	Attack (rifle, grenade)
		Strike/lockout/labor unrest	Pump failures	Construction, construction accidents
			Cooling water	Weather—flood/hurricane/tornado
			Major process reactor failures	Earthquake
			Major environmental releases (gas or water)	Civil disobedience

TABLE 7.2   Plant shutdown risk analysis table: Additional detail

Supply/storage problems	Exterior problems
Pipeline	Security—serious perimeter breach
Incoming storage tanks	Sabotage
Finished product storage tanks	Bomb
Pipeline	Explosions
Port/shipping problems	Attack (rifle, grenade)
	Construction, construction accidents
	Main plant maintenance
	Weather—flood/hurricane/tornado
	Earthquake
	Civil disobedience

    Spills (FPS)

    Fire (FPFr)

Port/shipping problems

    Dock problems (Dk)

    Shipping/arrival problems (SHAr)

    Loading/unloading problems (SHLd)

    Spills (SHSp)

    Sabotage (SABO)

    Pipeline/storage/port problems (SHPrt)

    Dock security issues (SHDkSec)

Exterior

    Security: Serious perimeter breach problems

        Unauthorized shipments in plant (EUs)

        Unauthorized vehicles in plant (EUv)

        Unauthorized personnel in plant (EUp)

        Personnel in unauthorized or restricted areas (EUpr)

        Suspicious packages (ESP)

        Weapons in the plant (EW)

        Sabotage: critical controls disabled (SCon)

        Spoofing of key sensor and control systems (SSpf)

        Cyber attack (SCA)

        Physical disabling of motor control centers (SMCC)

        Unauthorized or improper materials in process feed systems (SQA)

        Bomb: bomb or explosive device within the plant perimeter or nearby the fence line (B)

Explosions: explosions (EXP) or serious fires in adjacent properties (FIRE)

Attack: attack by a rifle, grenade, mortar, or other missile (ATK)

Construction

Construction activities during plant turnaround (CONTU)

New construction activities during normal operations (CONNO)

Construction accidents leading to fatalities or mass casualties (CONAC)

Significant damage to existing structures from new construction or repair activities (CONDAM)

Weather

Flood (FLD)

Tornado (TOR)

Heavy snow or ice (SNO)

Typhoon or hurricane (HUR)

Earthquake: Earthquake (EQ)

Tsunami (TSN)

Civil disobedience: Civil disobedience impacts or threatens the plant or plant personnel (CIVIL)

In each of these general categories, we have a number of subcategories, and any or all of these events could lead to a plant shutdown if conditions are right. The next step is to begin to categorize the likelihood and severity of the events so that we can get a general handle on the likelihood of their occurrence and form a general opinion on the degree of peril that the plant is in. And, of course, some of these events will come and go and have their own importance: for example, consider the earthquake, flood, or tsunami.

These incidents could be further classified according to their severity 1 and 2, where 1 is a significant event and 2 is a minor problem that could affect production.

Examples of the foregoing include:

Pipeline leak

Serious—requiring extensive repair and shutdown for a number of days

Minor—rupture repaired within 12 hours

Pipeline break

Large pipeline break ~30 cm or larger requiring days or weeks to repair

Smaller pipeline break ~15 cm or smaller requiring less than 2 days to repair

Pipeline corrosion

Significant corrosion requiring 10+ meters of repair/replacement

Minor corrosion requiring under 1 m of repair or replacement

Pumping station leak

An event that generates a leak causing significant contamination of the environment of over YY cubic meters and requiring a cleanup costing more than $XX

A minor spill or leak causing a leak or environmental contamination less than YY cubic meters and/or costing less than $XX for cleanup

Even with the simplest categorization using the 1s and 2s for severity, we now have a single branch of a fault tree that has five levels and close to 60 branches and 120 conditions that should be evaluated for severity and probability. By comparison, the BP Deepwater Horizon Spill Fault Tree Investigation[9] has four principal areas of investigation and several hundred individual potential causes. It is worthy of an examination.

FAILURE MODES AND EFFECTS ANALYSIS

Failure Modes and Effects Analysis (FMEA) is very similar to FTA, and it also contains some of the elements of bow-tie analysis: the analysis focuses on what can go wrong and how it can go wrong.

In setting up an FMEA, it is best to start small, with individual systems or subsystems, and not get too large because the size introduces complexity into the analysis and there is likelihood that the FMEA effort will get derailed or lost in the details and not reach a satisfactory conclusion. An FMEA is a dedicated, interdisciplinary team effort. The different disciplines are necessary to consider all the various failure modes.

The first step starts with the process selection. The process must be small enough to be manageable and have few enough variables to prevent it from becoming overwhelming. A typical security-related process might be the evaluation of the fencing and alarm systems, or the security of the dock facilities, or the security of the external pipeline systems rather than the internal pipeline systems. The reason for this selection of the latter category is the ease of identification and purpose, and in all likelihood, the pipeline will not get involved in identifying it with plant piping that has various sizes and various starting and end points.

Next, select the team for evaluation. In the example of the pipeline, the team should include designers, maintenance personnel, security personnel, the IT department, and the security department as well as someone from finance and accounting. While the latter may not be directly useful in the FMEA, they should be able to provide input as to the likelihood that financial resources and information would be available to implement solutions and to evaluate the potential costs of failures of specific magnitudes. In this instance, it might also be useful to involve the construction department (different from maintenance) because they can provide input on what may be required to facilitate repairs when needed. The team should also include a secretary (one of the team members or an outside consultant) who takes notes and has the technical background to understand the issues and record them properly for evaluation and record keeping.

Next, diagram all the steps in the process. Work out a flow sheet and identify the steps sequentially so that they can be easily referred to in the analysis phase. This task is not as easy as it may first appear. The final diagram for analysis should include all the inputs and control systems. The chemical engineering community refers to this as a process and instrumentation drawing, or a P&ID, and that is a good starting place. But it should be supplemented and simplified so that materials entering the process and their quantities and sources are identified. The outputs should be similarly identified. The final product should be agreed to by the team.

Following this, the drawing should be analyzed by the team to indicate all the areas where something can go wrong or where there may be a critical fault. The list should include minor and infrequent faults. This will generate a list of potential problem areas and hazards. This list should be tabulated. One type of appropriate analysis form is shown in Table 7.3.

The form will stretch on for a number of pages if properly executed. The team should identify the failure causes and effects. An arbitrary ranking on a scale of 1–10 should be used to indicate the severity of the incident and the likelihood of the occurrence and the likelihood of detection of the incident *before it occurs!* The same numerical scale for likelihood of occurrence, likelihood of detection, and severity of occurrence should be used, and the team agreement is a consensus activity. The risk profile number is obtained by multiplying the occurrence, detection, and severity numbers together, yielding a number between 1 and 1000.

The final column of action to reduce recurrence of failure should also be a consensus activity. There will be a number of items and actions, and it may be useful to record the solutions on separate sheets in a notebook, as there are sure to be much more than one solution for each item. The appropriateness of each solution should be highlighted, and the expansion or elucidation of the ideas should be cross-referenced to the form. The final report on the FMEA committee should be prepared and summarized in an indexed and clearly worded report with recommendations highlighted in bullet format with proposed costs and alternatives clearly defined.[10]

As an example, we have filled out the first two lines of the table to illustrate the types of factors to be considered.

In the first line, a simple sensor failure is providing a signal that indicates that there is a breach in the fence. The severity or criticality is low to moderate because there are other indicators and visual observation of the area in question.

In the second line, maintenance had removed a pump from a tank, but did not lock out and de-energize the electrical fittings. Maintenance also neglected to notify operations that the pump was out of service. The operator opened the valve and the spill happened. It could have been prevented by simple lockout–tagout procedures.

DHS ANALYSIS AND PLANS

The point of this section is to illustrate how the DHS has adopted the general principles involved in development of a security plan and to briefly outline the elements that comprise their requirements for security. Their approach to chemical plant security is entirely consistent with good practice as outlined in this book.

TABLE 7.3 FMEA worksheet (more extensive forms are available for free download from ASQ.org)

Incident number	Type of failure	Failure condition	Effects of failure	Probability of occurrence	Probability of timely detection	Severity	Risk profile number	Actions to reduce risk
1	Alarm	False alarm	Indicated security breach in fence	Frequent, daily until repair is made	Detection within 10 minutes	Important	6—false alarms	Fencing and sensing repair
2	Pump and valve	Failed open	Loss of 55,000 barrels of product	Rare but should be preventable	Shift change (8 hours)	Critical	12—plant spill	Institute lockout/tagout program to prevent recurrence
3								
4								
5								

After the attack on the World Trade Center on September 11, 2001 (9/11/2001), the US DHS was formed on September 22, 2001, and, with the passage of the Homeland Security Act of 2002, went into high gear with regard to all types of security systems in the United States and particularly the chemical industry sector. Their program is the Chemical Facility Anti-Terrorism Standards found in the US Code of Federal Regulations 6 CFR 27.215 and 6 CFR 27.235. The program is essentially divided into four tiers based upon the type of chemicals and the volume of chemicals produced as listed in 6 CFR Part 27 (FR Vol. 72, No. 223, Tuesday November 20, 2007, pp. 65421–65435). This is known as the Top Screen Process. Any company having an inventory greater than the screening threshold will fall into tier 1, 2, 3, or 4. Tier 1, 2, or 3 facilities must develop and submit a security plan along the following guidelines. Starting at 6 CFR 27.225, the risk-based security standards are to be implemented as follows:

§27.225 site security plans

(a) The site security plan must meet the following standards:

(1) Address each vulnerability identified in the facility's security vulnerability assessment, and identify and describe the security measures to address each such vulnerability.

(2) Identify and describe how security measures selected by the facility will address the applicable risk-based performance standards and potential modes of terrorist attack including, as applicable, vehicle-borne explosive devices, waterborne explosive devices, ground assault, or other modes or potential modes identified by the department.

(3) Identify and describe how security measures selected and utilized by the facility will meet or exceed each applicable performance standard for the appropriate risk-based tier for the facility.

(4) Specify other information the assistant secretary deems necessary regarding chemical facility security.

(b) Except as provided in §27.235, a covered facility must complete the site security plan through the CSAT process or through any other methodology or process identified or issued by the assistant secretary.

(c) Covered facilities must submit a site security plan to the department in accordance with the schedule provided in §27.210.

(d) Updates and revisions:

(1) When a covered facility updates, revises, or otherwise alters its security vulnerability assessment pursuant to §27.215(d), the covered facility shall make corresponding changes to its site security plan.

(2) A covered facility must also update and revise its site security plan in accordance with the schedule in §27.210.

(e) A covered facility must conduct an annual audit of its compliance with its site security plan.

§27.230 Risk-based performance standards

(a) Covered facilities must satisfy the performance standards identified in this section. The assistant secretary will issue guidance on the application of these standards to risk-based tiers of covered facilities, and the acceptable layering of measures used to meet these standards will vary by risk-based tier. Each covered facility must select, develop in their site security plan, and implement appropriately risk-based measures designed to satisfy the following performance standards:

(1) Restrict area perimeter. Secure and monitor the perimeter of the facility.

(2) Secure site assets. Secure and monitor restricted areas or potentially critical targets within the facility.

(3) Screen and control access. Control access to the facility and to restricted areas within the facility by screening and/or inspecting individuals and vehicles as they enter, including:

(i) Measures to deter the unauthorized introduction of dangerous substances and devices that may facilitate an attack or actions having serious negative consequences for the population surrounding the facility

(ii) Measures implementing a regularly updated identification system that checks the identification of facility personnel and other persons seeking access to the facility and that discourages abuse through established disciplinary measures

(4) Deter, detect, and delay. Deter, detect, and delay an attack, creating sufficient time between detection of an attack and the point at which the attack becomes successful, including measures to:

(i) Deter vehicles from penetrating the facility perimeter, gaining unauthorized access to restricted areas or otherwise presenting a hazard to potentially critical targets

(ii) Deter attacks through visible, professional, well-maintained security measures and systems, including security personnel, detection systems, barriers and barricades, and hardened or reduced value targets

(iii) Detect attacks at early stages, through countersurveillance, frustration of opportunity to observe potential targets, surveillance and sensing systems, and barriers and barricades

(iv) Delay an attack for a sufficient period of time so to allow appropriate response through on-site security response, barriers and barricades, hardened targets, and well-coordinated response planning

(5) Shipping, receipt, and storage. Secure and monitor the shipping, receipt, and storage of hazardous materials for the facility.

(6) Theft and diversion. Deter theft or diversion of potentially dangerous chemicals.

(7) Sabotage. Deter insider sabotage.

(8) Cyber. Deter cyber sabotage, including by preventing unauthorized on-site or remote access to critical process controls, such as supervisory control and data acquisition (SCADA) systems, distributed control systems (DCS), process control systems (PCS), industrial control systems (ICS), critical business system, and other sensitive computerized systems.

(9) Response. Develop and exercise an emergency plan to respond to security incidents internally and with assistance of local law enforcement and first responders.

(10) Monitoring. Maintain effective monitoring, communications, and warning systems, including:

(i) Measures designed to ensure that security systems and equipment are in good working order and inspected, tested, calibrated, and otherwise maintained

(ii) Measures designed to regularly test security systems, note deficiencies, correct for detected deficiencies, and record results so that they are available for inspection by the department

(iii) Measures to allow the facility to promptly identify and respond to security system and equipment failures or malfunctions

(11) Training. Ensure proper security training, exercises, and drills of facility personnel.

(12) Personnel surety. Perform appropriate background checks on and ensure appropriate credentials for facility personnel and, as appropriate, for unescorted visitors with access to restricted areas or critical assets, including:

(i) Measures designed to verify and validate identity

(ii) Measures designed to check criminal history

(iii) Measures designed to verify and validate legal authorization to work

(iv) Measures designed to identify people with terrorist ties

(13) Elevated threats. Escalate the level of protective measures for periods of elevated threat.

(14) Specific threats, vulnerabilities, or risks. Address specific threats, vulnerabilities, or risks identified by the assistant secretary for the particular facility at issue.

(15) Reporting of significant security incidents. Report significant security incidents to the department and to local law enforcement officials.

(16) Significant security incidents and suspicious activities. Identify, investigate, report, and maintain records of significant security incidents and suspicious activities in or near the site.

(17) Officials and organization. Establish official(s) and an organization responsible for security and for compliance with these standards.

(18) Records. Maintain appropriate records.

(19) Address any additional performance standards the assistant secretary may specify.

Facilities that fall into the tier 4 requirements can submit an alternative security plan. The plans are shared with the DHS and reviewed by DHS, plus the required annual review by the company subject to the regulations.

BOW-TIE ANALYSIS

Bow-tie analysis is a graphical way of hazard or incident assessment that is applicable to specific processes. It does not lend itself well to more broad security and risk analysis, but it does have the advantage of illustrating not only the hazards and initiating events but the recovery steps as well.

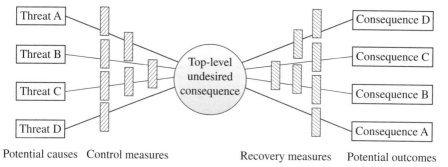

Figure 7.3   Example of bow-tie analysis.

The risk assessment or hazard assessment has been examined and discussed earlier. The bow-tie method is illustrated in Figure 7.3:

The bow-tie method figure above is somewhat self-explanatory. The unwanted event is placed in the center of the diagram, and threats, causes, and attacks are on the far left. Moving to the right are the control or mitigation measures, which form an upper and lower bound on the attack and tend to prevent its occurrence. On the right of the event are the recovery measures and the potential consequences. It is possible for one control or recovery measure to span more than one threat and for one recovery measure to have more than one outcome. The advantage is that the "cause and the fix" are on one format and can be compared. The disadvantage is that the level of detail about the attack and recovery may be primarily in summary form and not have sufficient detail to provide meaningful input. The detail may require supplemental sheets or plans.[11]

Example

In the figure above, let us assume that there is a large tank of petroleum (flammable) material, and a tank of highly odorous trimethyl amine nearby. The consequences from a petroleum leak (threat A) would be fire, smoke, vapor cloud, and possibly community evacuation. Trimethyl amine is a highly odorous compound that smells like dead fish.

The nose can detect concentrations of TMA as low as $0.267\,mg/m^3$ in air, and the compound has a high vapor pressure ($1189\,mm\,Hg\,@\ 15°C$, while water is $187.5\,mm\,Hg\,@\ 15°C$) and will evaporate out of water, and it has a recommended exposure limit (8 hours TWA between 10 and $15\,mg/m^3$), such that it will outgas from water. A release of the material can be both irritating and hazardous to the health of the community.[12] A release should be avoided at almost all costs.

Top-level undesired consequences are:

1. Community evacuation

2. Fire

3. Explosion

4. Community exposure to chemicals

Threat scenario A is a leak in a petroleum tank.
Threat scenario B is a leak in any part of the TMA plant.

Control measures for A:

Continuous inspection of tanks and piping

Periodic pressure testing of lines

Periodic maintenance of seals and pumps

Electrical grounding of all hoses and piping

Fire foam for dikes around the tank

Control measures for B:

Periodic inspection and testing of all valves and joints with electronic sniffers to detect leaks

Use of welded steel pipes, minimizing flanges

Regular maintenance

Grounding of all piping

Periodic replacement of pump seals

Enclose plant within a structure

Use of acid scrubber for gases from the building to reduce TMA

Recovery measures for A:

Diking for spill control

Vapor blanketing for diked area

Spill control drills

Rapid response for cleanup

Firefighting measures

Recovery measures for B:

Community evacuation

Shelter in place

Acid water spray to attempt to knock down vapor cloud

Potential consequences for A:

Fire

Underground contamination that needs to be cleaned up

Expensive spill cleanup

Loss of product

Potential consequences for B:

Adverse long-term health effects

Bad community relations

Expensive decontamination of residences and properties

HAZOPS AND PROCESS SAFETY MANAGEMENT

HAZOPS is a program that is designed to deal with and assess hazardous chemical production and overall chemical safety in manufacturing and handling operations. The program is specified under US Regulations in 29 CFR (OSHA). There are several approaches to HAZOPS. The first approach is fault tree, which we have already covered. A second approach to HAZOPS is the "what if?"; a third approach is the checklist approach. Each is designed to identify process safety hazards that would include individual worker safety issues and overall plant and production safety issues in manufacturing, storage, handling, shipping, and chemical reactivity.

Some in the operations group of a plant would tend to argue that HAZOPS and PSM are the exclusive property of the engineering, safety, and management departments. However, there are a number of ways in which security has to be involved in the response to any and all incidents. Therefore, they should be included in any analyses.

The overall HAZOPS process requires a lot of information on process safety. Note specifically that the information required for an operator of a chemical facility is a complete documentation of the process, all pertinent records on reactions and corrosion, etc. and information on control limits, how to start and shut down the process, and how to restore order to the process once it gets out of control.

The entire regulations can be found in 29 CFR 1910.119 and 1910.109 under process hazard analysis (PHA) at the reference cited at the end of this paragraph.[13]

OSHA also references a number of relevant training documents that can be helpful for analysis. One is a power point document that outlines the basics of the PSM process, and the second is the checklist for auditing PSM compliance.[14] A brief outline of the critical elements of the PSM process and their applicability to security concerns is as follows.

Process safety information: General

Employers must complete a compilation of written process safety information before conducting any PHA required by the standard. The compilation of written process safety information, completed under the same schedule required for process hazard analyses, will help the employer and the employees involved in operating the process to identify and understand the hazards posed by those processes involving highly hazardous chemicals. Process safety information must include information on the hazards of the highly hazardous chemicals used or produced by the process, information on the technology of the process, and information on the equipment in the process.

The approach here is to indicate the relative information for planning processes as it may relate to physical security of the plant. In the master file of the plant, available to plant security and to select management teams in the security force, the following information should be included in the emergency and security plans at the plant, where appropriate. The purpose is to provide enough information for emergency responders and guards to facilitate evacuations and/or work safely when there may be an incident in the plant.

In that regard, a planning tool such as ALOHA[15] is excellent for planning purposes because it permits the evaluation of the on and off plant effects of accidental chemical releases.

PHA and HAZOPS

The PHA and HAZOPS are a process that is applicable to the process and manufacturing industries. It is primarily designed as a safety analysis tool for plant operations. There are several different ways of performing PHA process as it may apply to plant security. The applicable methods include a review of a detailed checklist prepared with regard to the process and a comprehensive analysis of the modes of failure and restoration: a "what-if" analysis that reviews the plant processes literally pipe by pipe and process, looking for methods of what can happen and/or go wrong.

For the security professional, PHA and HAZOPS have some of the same elements but with a slightly different focus. We will designate this by the *SPHA*.

A SPHA study identifies hazards and operability problems by identifying how the plant security system might deviate from the design intent. If a solution to a problem becomes apparent, it is recorded as part of the SPHA result, but the prime objective for the SPHA is problem identification.

SPHA studies are normally conducted during the design phase, especially when new technology or process are involved as part of the larger PHA in the plant, but can be used at almost any phase of a plant's or security system's life.

The PHA and SPHA is based on the principle that several experts with different backgrounds can interact and identify more problems when working together than when working separately and then combining their results. The most common form of HAZOPS study employs guide words to test the consequences of parameters deviating from design.

The objectives of an SPHA study may include as follows: (i) check the safety of a design or perimeter or remote location; (ii) check the maintainability and operability of a design or configuration and equipment; (iii) decide whether and where to build additional security measures or improvements; (iv) develop a list of questions to ask a supplier of equipment; (v) check operating and safety procedures, and test the security of an operation or part of the plant.

Additional purposes may include improvement of the safety of an existing facility through examining various elements of the security, fire, health, safety, and other divisions and their coordinated functioning in time of a plant-wide incident and verification that security instrumentation is reacting to optimum parameters and has minimized interferences.

Consequences to be considered   It is also important to define what specific consequences are to be considered if there is a security breach or nonperformance and the way it may affect employee safety, loss of plant or equipment, loss of production, liability, insurability, public safety and impact on the neighborhood surrounding the plant, and, if appropriate, environmental impacts due to incidents arising from security failures or responding to incidents dealing with plant fire, explosion, or other causes.

The SPHA team must be chosen from experienced people preferably with knowledge of a similar facility who will likely be involved with the operation of the plant and someone who is intimately familiar with the functioning of the security operations at the facility. The team leader should be chosen for his/her ability to get the team to focus on making the analysis rather than the ability to solve problems. The issues identified can be resolved after the SPHA. Depending on the scope of the SPHA, the following team assignment is suggested.

HAZOPS- or SPHA-specific guide words—simple words used to qualify or quantify the intention in order to guide and stimulate the brainstorming process and so discover deviations or failures or problems in the security function.

Note: HAZOPS reviews use specific guide words that may or may not be applicable to the security function. The use of HAZOPS guide words is suggested and may be helpful, but not mandatory because of the specialized purpose of the SPHA.

The success or failure of the SPHA review depends on several factors: (i) the completeness and accuracy of drawings and other data used as a basis for the study; (ii) the technical skills and insights of the team; (iii) the ability of the team to use the approach as an aid to their imagination in visualizing deviations, causes, and consequences; and (iv) the ability of the team to prioritize and concentrate on the more serious failures or flaws that are identified.

The following checklists are for PHA at the plant level, but plant security should play a role in the overall planning. Table 7.4 is an example of the PHA process and the security's role in the PHA.

ALOHA, CAMEO, AND SECURITY PLANNING TOOLS

The CAMEO suite is for computer-aided management of emergency operations. It is available free of charge as a download from the US Environmental Protection Agency. The website is http://www2.epa.gov/cameo/what-cameo-software-suite. It consists of several tools that are useful in planning for emergencies. CAMEO was developed from ARCHIE, an old and now out-of-date computer program developed for emergency response. The CAMEO suite includes several programs.

The CAMEO database includes a database and information management tool that will assist US facilities in the preparation of their data management and reporting requirements under the Emergency Planning and Community Right to Know Act, which affects US facilities only.

CAMEO chemicals includes chemical response datasheets and a reactivity prediction tool that has UN/NA datasheets providing information on health hazards, physical properties, air and water hazards, spill response, and firefighting recommendations. The information base for much of the material is from the Emergency Response Guidebook and the Hazardous Materials Table (49CFR172.101). The program also allows one to mix chemicals and predict the reactions (http://www.cameochemicals.noaa.gov). The file is about 43 meg and is available for handheld computers and is accessible online.

MARPLOT is a mapping application. Because the CAMEO suite was developed by the USEPA, the mapping database accompanying the program is primarily

TABLE 7.4 Process hazard analysis and security's role

Basic physical and chemical information that should be available: Toxicity Permissible exposure limits Physical data of chemicals and equipment Reactivity data Corrosivity data Thermal and chemical stability data Hazardous effects of inadvertent mixing of different materials	This information should be applicable to the guard force in a reference manual or tables so that the planning exercises can be conducted using realistic scenarios for disaster mitigation
A block flow diagram or simplified process flow diagram	The block flow diagram should show the locations of critical shutoff valves and fire extinguishers, sewer covers, and spill kits
Process chemistry Maximum intended inventory Safe upper and lower limits for such items as temperatures, pressures, flows, or compositions An evaluation of the consequences of deviations, including those affecting the safety and health of employees	The information on deviations should be summarized to inform the guard force about the type of incident and its probable cause and location
Material safety data sheets	In any event and in multiple places around the facility, MSDS should be posted or otherwise available to all plant employees and the guard force that would have multichemical and multilocation exposure during an incident The guard force should also be included in training for first responders and where necessary be instructed and trained with safety suits and equipment
Process information	This information should be summarized according to the potential major hazards. For example, high voltage, volatile chemicals, acids, bases, reactive materials, etc. should also be identified
Storage tanks	Storage tanks should be identified by type, contents, diking capacity, and drainage system, including hazards

In 1975, the Gulf Oil Refinery, near Philadelphia, PA, had a tank fire. Firefighters were standing in an adjacent diked area spraying water and fire foam on the burning tank and did not recognize that they were standing in water that had a volatile layer of oil floating on top of it. The fire foam provided an insulating blanket preventing volatilization of the floating oil. When the blanket was ruptured (cause unknown), the volatile vapors flashed and caught fire catastrophically killing eight firefighters and burning up a fire truck. See the description in Wikipedia: http://en.wikipedia.org/wiki/1975_Philadelphia_Refinery_Fire

Ventilation systems and relief systems and general description of safety systems

Necessary for plant safety during an incident and should indicate the type of shutoff controls and their locations

The plant should have a periodic safety review by one or more of the following techniques:

- What if
- Checklist
- What if/checklist
- Hazard and operability study (HAZOP)
- Failure mode and effects analysis (FMEA)
- Fault tree analysis
- An appropriate equivalent methodology

These are methods of assessing the safety and security of the plant and its operations

Operating procedures

Required for guard force and security operations. This does not necessarily include the guidance for the operators but should be a well-thought-out set of routine and emergency instructions to the guard force

Note: The instructions should not just simply say: "Contact the Dispatch or control center for instructions," but should be detailed enough to permit the guard force to participate in the resolution of the emergency

Written action plans and training

While the OSHA PSM standard applies more thoroughly to the plant, a comparable level of planning and training should be given to the guard force with thoughts toward their utilization during an incident whether it be a security breach or a major plant conflagration

Refresher training

The training should be updated annually and should be thoroughly documented

Security procedures should include contract labor, as well as new employees

Contractors in the plant may represent a unique hazard, as they are not vetted to the same extent as plant employees

New employees, especially those in critical positions, should be copilot trained until their activities and reliability can be assured

The procedure for this is described in one of the award winning master's theses from the Naval Postgraduate School Center for Homeland Defense and Security. The program consists of copiloting new employees with different experienced and trusted personnel who will also conduct their reviews and evaluations. The program was titled "No Open Doors"

(Continued)

TABLE 7.4 *(Continued)*

Permit enforcement within the plant	Items such as hot work, lockout/tagout, confined space, etc. are critical to safe operation of any facility. The question of who issues and enforces the permits must be addressed with plant security as well as operations. Normally, the security force is not included in this operation
Incident investigation	A member of the security force should be on the team that investigates all plant incidents and accidents
Shipping and receiving (trade)	The security force should be heavily involved in monitoring warehouse and shipping and receiving operations, not only to prevent theft, but also for vessel and vehicle security
Normally unoccupied and remote facilities	This represents a special challenge to the security force as well as to operations because the possibility of sabotage or outside attack on these facilities is low risk for the attacker and has a high probability of success
Dangerous goods/radioactives/and other storage areas	While these are routinely a part of the warehouse operations, special precautions may be required for storage areas, magazines, and other facilities that handle highly energetic compounds and/or radioactive materials because of the specialty licensing requirements
Waste disposal	Security should also be involved in waste disposal operations as a precaution, not only to prevent loss, but also to prevent acts of sabotage or explosives from being smuggled into the plant
Emergency response operations	During a plant incident, the various roles of fire, emergency, ambulance, other support and external services, as well as internal plant operation departments are called into service. Plant security plays a critical role in this emergency, providing coordination and control services, admitting personnel to the plant, excluding unwanted or unnecessary personnel, and communicating with outside services that may be required to provide an evacuation of the facility or the community. These areas should be addressed in the plan

for US cities, but it is applicable worldwide. The MARPLOT program allows the user to enter local mapping that may include schools, property lines, etc. and then see the generated data.

ALOHA is the Areal Locations of Hazardous Atmospheres, and it includes dispersion modeling programs that the user can control. In order to use the program, one needs to know something about air dispersion modeling. The user can control atmospheric conditions for the modeling, including horizontal mixing. It will also model dense gas dispersion. The user can, within very broad limits, set and control the materials of release and the type of release, and the program will model fireballs, gas releases, etc. One can set the type of tank, whether it is full or empty, horizontal or vertical, under pressure, and partially full or full. The program will allow the user to map the hazard areas associated with the release. It is designed to provide emergency information for planning purposes and first responders. We have used this in a large number of applications for planning purposes and have found it excellent![16] The program models single- and two-story buildings and allows the user to set the number of air changes per hour in the building.

THE COLORED BOOKS

The EU has come up with a number of manuals on reliability and security. These were prepared under the auspices of the European Union Economic Commission for Europe (EUECE).

The colored books are so named by the color of their jackets.

The first of the colored books really has no color at all. It was prepared by the EUECE and is titled as follows.

Generic Guideline for the Calculation of Risk Inherent in the Carriage of Dangerous Goods by Rail

This guideline and as such it includes a discussion of the movement of dangerous goods by rail, and it has a good discussion on the subject of risk assessment of rail movement and accident rates for transportation incidents.

The Orange Book: Management of Risk—Principles and Concepts

The Orange Book is accurately titled with regard to its contents. It is relatively short weighing in at 52 pages, but it discusses all of the major concepts involved in risk management, although not in the detail covered in this book. The book is published by the Royal Treasury in the United Kingdom and is available for free download: https://www.gov.uk/government/publications/orange-book.

The opening lines of the Orange Book recommend that it be read in conjunction with the Green Book (which will be discussed in following paragraphs). The Orange Book lays out the general principles of risk management and has several unique observations:

- "The management of risk is not a linear process; rather it is the balancing of a number of interwoven elements which have to be in balance with each other if risk management is to be effective. Furthermore, specific risks cannot be addressed in isolation from each other; the management of one risk may have an impact on another, or management actions which are effective in controlling more than one risk simultaneously may be achievable."

- "There is no single right way to document an organization's risk profile, but documentation is critical to effective management of risk."

- There are two different phases of risk identification:

  ◦ *Initial risk identification* (for an organization that has not previously identified its risks in a structured way, or for a new organization, or perhaps for a new project or activity within an organization).

  ◦ *Continuous risk identification* is necessary to identify new risks that did not previously arise, changes in existing risks, or risks that did exist ceasing to be relevant to the organization (this should be a routine element of the conduct of business).

- Risks should be related to objectives.

- "Individual risks which an organization identifies will not be independent of each other; rather they will typically for natural groupings…. It is important not to confuse the grouping of risks with the risks themselves…. All risks once identified, should be assigned to an owner who has responsibility for ensuring that the risk is managed and monitored over time."

- Important risk assessment principles include:

  ◦ Insuring that a clearly structured process identifies both the likelihood and the impact for each considered risk

  ◦ Recording and documenting risk considerations in a manner that can facilitate identification and monitoring of the risk priorities

  ◦ Clarity in describing the difference between inherent and residual risk[17]

- Risk management can be dealt with by (i) toleration, (ii) treatment, (iii) transferring, or (iv) termination of the risk elements.

- Recommendations include monitoring of the risk profile, performing risk review annually, and making provisions to alert management to changing (increasing) levels of risk.

- Much of the risk assessment is focused on financial risk (where it needs to be), but the principles are primarily relating to those dealing with accounting practices, and the document relies on Risk Management Guidance and the Mullarkey Report (2003) on risk assessment that is generally unavailable because it has been institutionalized.

- A copy of parts of the Mullarkey Report is available from the Government of Ireland, Government Accounting Section, Department of Finance, March 2004, *Risk Management Guidance for Government Department and Offices.*

- The guidance document cited in the preceding paragraph uses a *risk register,* which is a guidance document summarizing risk including categories for a brief description of the risk item; the division to which it is assigned; the policy or strategy number; a numeric ranking for likelihood, impact, and control effectiveness; a rating number that is obtained by multiplying the three rankings together; a brief statement of consequences; measures to address; additional actions; and finally the owner of the issue (person responsible for minimizing the risk). In short, it similar to the type of table used in a bow-tie analysis that is pinned to a division and an owner of the risk.

- One other document used in the United Kingdom is the HM Treasury *Risk Management Assessment Framework: A Tool for Departments* (July 2009). The document takes a slightly different approach to the Mullarkey principles but provides a comprehensive and thorough discussion for review.

The Green Book: Methods for the Determination of Possible Damage to People and Objects Resulting from Release of Hazardous Materials, CPR-16E

This document has seven chapters in 337 pages. The chapters address (1) damage caused by heat radiation, (2) the consequences of explosion effects on structures, (3) the consequences of explosion effects on humans, (4) survey study of the products that can be released during a fire, (5) damage caused by acute intoxication, (6) protection against toxic substances by remaining indoors, and (7) population data.

Chapter 1 of the Green Book deals with the effects of heat radiation and has useful curves and data and formulas on burn/exposure data for exposed skin and clothing ignition from various sources. The first chapter on radiation damage also considers the radiation values for ignition of building materials, softening steel members, and the formulas for the evaluation of these cases.

The limiting value for third-degree burns varies between 125 and 140 kJ/m², for durations of 1–35 seconds. For second-degree burns, the minimum value is between 30 kJ/m² at about 3 seconds and about 75 kJ/m² at 30 seconds for guidance. The formula for determining the degree of burn based on a heat exposure and time is complex and involves error function curves and higher mathematics. There is also a good discussion on pool fires and fireball (BLEVE) explosions. As such, it is useful for planning purposes but extremely complex and perhaps unusable in an emergency situation without substantial evaluation and applications.

Chapters 2 and 3 of the Green Book deal with blast effects and loadings, are excellent, and have many useful tables and graphs on the calculation of blast effects on a building and upon humans. There is also a substantial section on the extremely complex subject of the building response to blast and shock loadings. The analysis is quite detailed on the subject of natural periodicity of buildings in response to shock loadings.

There is also a section on the empirical data of blast loadings and effects: on buildings, pressure loadings on the order of the following will create significant damage.

Damage level	psi	kPa
Total destruction of structure	>12	>83
Heavy damage	>5	>35
Moderate damage	>2.5	>17
Minor damage	>0.5	>3.5

There is also a useful gauge that compares the relative damage levels and difference between English- and US-built homes, where the English homes are generally brick built, and it generally follows the information outlined immediately above, but it is interesting to observe that comparable low level pressure events on US-style homes appear to cause less damage than in brick-built homes. This is believed due to the difference in building materials where US homes tend to use more wood. The appendices to the chapter on the effects of explosions on buildings contain a number of good models for additional mathematical analysis.

Chapter 3 of the Green Book principally deals with the effects of blasts on human populations, including the effects of explosions and flying debris such as glass on the body and on the skin. Part of the chapter deals with the crushing of organs and lungs due to the explosive force and internal damage of people in the vicinity of an explosion. Overall, the material is quite detailed and quite useful.

Chapter 4 of the Green Book concerns itself with the effects of damage caused by combustion products and particularly those that have organics and chlorine compounds. As such, there is a good bit of information on the formulation of combustion products that result in the formation of polychlordibenzo-*p*-dioxins and other polychlorinated compounds. The information on combustion by-products is both useful and timely with regard to planning for any incidents.

Chapter 5 of the Green Book deals with the effects of acute intoxication. But the title is misleading. The chapter does not deal with alcoholism but deals with inhalation of toxic substances. As such, it provides an excellent guide to the subject of inhalation. It is not as comprehensive as the NIOSH Pocket Guide to Hazardous Chemicals, but it is good.[18] The personal preference is for the NIOSH guide because it is both available on PDF and is designed for handheld computers.

Chapter 6 of the Green Book discusses the protection against toxic substances by remaining indoors. The relative effect of protection is dependent upon a number of factors, including wind speed, concentration in the toxics cloud, insulation of the building, ventilation rate of the structure, absorption of the hazardous materials, particle or molecule size, and other factors. It cites the idea that most residential structures have a ventilation rate of $0-0.5^{-h}$ so that the protection factor can be preliminarily assessed as equal to 2.0 for a hazardous emission duration of 1 hour or less. The detailed information is presented in a form that will enable the sheltering capabilities of residences and structures. While this information is useful in the planning stage, it may be too complex to use in an emergency response situation, and much of the information is covered in the ALOHA program. The chapter does contain a series of very good graphs indicating concentration reduction curves from a temporary source with relative ventilation rates for residential and other buildings and tables that allow one to calculate their relative protection factors from various types of releases.

Chapter 7 of the Green Book presents planning data on population densities. The information is useful, but a cautionary note should be sounded with regard to different densities in various countries. In the United States and Canada, land use planning data are available, and they may provide a more accurate guide for the existing populations in the vicinity of a plant or facility. The data in Chapter 7 is excellent but of necessity in general. Land use planning data from various governmental agencies may provide more accurate pictures of local population densities and is certain to provide a more recent picture that is influenced by regional differences.

The Yellow Book: Methods for the Calculation of Physical Effects due to the Releases of Hazardous Materials (Liquids and Gases), CPR-14E

This book is massive at 870 pages but fortunately available in PDF. The opening paragraph in user instructions sets the tone for the book:

> The educational design provides a framework according to which this version of the Yellow Book has been structured…the Yellow Book starts with a section on outflow and spray release (Chapter 2), then addresses evaporation (Chapter 3) and dispersion (Chapter 4), before addressing several other specific aspects, such as vapor cloud explosion (Chapter 5), heat load (Chapter 6), and rupture of vessels (Chapter 7). Finally a section on interfacing related models (Chapter 8) illustrates how to proceed in applying a sequence of models in estimating physical effects according to a few selected scenarios.

The Yellow Book goes into numerical modeling of various types of releases, including sprays and vapors, and has advice on the topics of dense gas modeling and modeling evaporation from a pool and a pool fire. It is the type of book that a chemical engineer uses to calculate mathematical effects, and one needs a sound foundation in heat transfer, thermodynamics, engineering, and some familiarity with atmospheric dispersion modeling in order to use this book properly. It is literally overwhelming in the amount of details provided and required to create models of the various types of releases. Of particular interest is Chapter 7, which deals with tank ruptures and blasts. This chapter also addresses the subject of blast projectiles and tank fragments that might be released during a tank rupture scenario.

The Red Book: Methods for Determining and Processing Probabilities, CPR-12

This is really a book about applied statistics and risk assessment. As such, it covers many of the topics addressed in this book but in significantly greater mathematical detail. It addresses topics such as mean time between failures (MTBF), FTA and event tree analysis, Markov chains, accident sequence development and quantification, as well as detailed discussions on statistics and computation and reliability theory. Of particular interest are Chapters 16 and 17, which deal with reliability availability maintenance and reliability-centered maintenance, with practical guides to establishing the mathematical basis for the programs and suggestions for establishing tenders for the program if conducted by outsiders. Overall, the book is excellent and informative.

The Purple Book: Guidelines for Quantitative Risk Assessment, PGS 3

Special note: Many industries may encounter problems in applying the quantitative risk assessment procedures because the focus of the assessments ultimately is in fatalities per year. (This is often referred to as the probit function.) From a liability standpoint, US-based corporate executive would not consider designing or evaluating a system in terms of the number of deaths per year because the aggressive legal culture in the United States would indict him and his company on criminal negligence charges.

The analysis of the Purple Book is a bit more comprehensive than the other colored books because of its relevance to the overall topics of security. The approach taken in the Purple Book is slightly different from the standpoint of risk assessment. Performing a quantitative risk assessment is a lot of work and requires detailed analysis of the entire facility, and it could require several hundred to several thousand man-hours of effort. The EUECE and the UK Competent Authority (Health and Safety Executive in the United Kingdom) have developed a screening test to determine whether or not a specific facility would require a comprehensive QRA.

The need for the QRA is based on a screening value S derived by computing the boundary distances between the operating portions of the plant and the plant fence line. The guideline values are determined by the following set of calculations.

First, criteria are set for excluding substances from the risk assessment guidelines to reduce the potentially large number of sites:

1. *Physical form of the substance*

    Substances in solid form such that under both normal conditions and any abnormal conditions that can be reasonably foreseen, a release of matter or of energy, which could create a major accident hazard, is not possible

2. *Containment and quantities*

    Substances packaged or contained in such a fashion and in such quantities that the maximum release possible under any circumstances cannot create a major accident hazard

3. *Location and quantities*

    Substances present in such quantities and at such distances from other dangerous substances (at the establishment or elsewhere) that they can neither create a major accident hazard by themselves nor initiate a major accident involving other dangerous substances

4. *Classification*

    Substances that are defined as dangerous by virtue of their generic classification in Annex I, Part 2, of Council Directive 96/82/EC but that cannot create a major accident hazard and for which therefore the generic classification is inappropriate for this purpose

The next step is to subdivide the installation or plant into subunits and calculate an A number for the subunit. The number A is calculated by

$$A = Q \times \frac{O_1 \times O_2 \times O_3}{G}$$

where Q is the quantity of the substance present from a list in Section 2.3.2.1 of the document, O_1–O_3 are factors for process conditions described in Section 2.3.2.2, and G is the limiting quantity as described in Section 2.3.2.3.

The O_1 factor is either 1 or 0.1 depending if the quantity is in process or in storage.

The O_2 factor is defined in the following table:

Positioning	O_2
Outdoor installation	1.0
Enclosed installation	0.1
Installation situated in a bund and a process temperature T_p Less than the atmospheric boiling point T_{bp} plus 5°C, that is, $T_p \leq T_{bp}+5°C$	0.1
Installation situated in a bund and a process temperature T_p More than the atmospheric boiling point T_{bp} plus 5°C, that is, $T_p > T_{bp}+5°C$	1.0

And the O_3 factor is dependent upon process conditions as follows:

Phase	O_3
Substance in gas phase	10
Substance in liquid phase	
Saturation pressure at process temperature of 3 bar or higher	10
Saturation pressure at process temperature of between 1 and 3 bar	$X+\Delta$
Saturation pressure at process temperature of less than 1 bar	$P_i+\Delta$
Substance in solid phase	0.1

$X = 4.5 \times$ saturation pressure (bar) $- 3.5$, and P_i is the partial pressure (bar) at the operating or processing temperature.

Δ is for liquids and is dependent upon boiling point of the liquid given below:

	Δ
$-25°C \leq T_{bp}$	0
$-75°C \leq T_{bp} < -25°C$	1
$-125°C \leq T_{bp} < -75°C$	2
$T_{bp} < -125°C$	3

Finally, the limit value G is based upon the toxicity of the materials as shown below:

LC_{50} (rat, in h, 1 h) (mg/m³)	Phase at 25°C	Limit value (kg)
$LC \leq 100$	Gas	3
	Liquid (L)	10
	Liquid (M)	30
	Liquid (H)	100
	Solid	300
$100 < LC \leq 500$	Gas	30
	Liquid (L)	100
	Liquid (M)	300

LC$_{50}$ (rat, in h, 1 h) (mg/m^3)	Phase at 25°C	Limit value (kg)
	Liquid (H)	1000
	Solid	3000
500<LC≤2000	Gas	300
	Liquid (L)	1000
	Liquid (M)	3000
	Liquid (H)	10,000
	Solid	∞
2000<LC≤20,000	Gas	3000
	Liquid (L)	10,000
	Liquid (M)	∞
	Liquid (H)	∞
	Solid	∞
LC>20,000	All phases	∞

With the additional caveat that flammable substances should not exceed a G value of 10,000 kg and explosive substances equal to or less than 1000 kg of TNT equivalent.

The A numbers are summed for different processes, toxics, explosives, and flammable substances, and then a screening number is developed at eight different locations around the perimeter of the facility. The screening number is of the form

$$S = \left(\frac{100}{L}\right)^N \times A$$

where N is 2 for toxics and N is 3 for flammables and explosives. L is the distance from the installation to the specific location in meters, with a minimum value of L = 100. The selection number has to be calculated for every installation at a minimum of eight locations on the boundary of the establishment, with a minimum distance of 50 m between each selection point. If the establishment has a boundary on the water, then the S value must be calculated on the bank side opposite the establishment.

Selection criteria for performing a QRA   The need for performing a comprehensive QRA is dependent upon the determined values for S at the facility. The QRA is required if:

The selection number of an installation is larger than one at a location on the boundary of the establishment (or on the bank side situated opposite the establishment) and larger than 50% of the maximum selection number at that location.

The selection number of an installation is larger than one at a location in the residential area, existing or planned, closest to the installation.

There are additional conditions specified for pipelines in the Purple Book that address gas and liquid pipelines and the hazard they can pose.

The Purple Book includes a number of data points for tanks and pipelines that are useful in determining the risk levels for loss of containment for industrial operations. Selected risk levels for loss of containment are shown below. For additional detail, see the Purple Book.

Loss of containment statistics for selected process equipment *Special note: The probabilities expressed below have a factor of safety already built in, and there is a caution in the Purple Book that the cumulative probability of loss of containment should never exceed 10^{-6} per year.*[19]

	Tanks and vessels		
	Instantaneous	Continuous 10 minutes	Continuous >10 minutes
Pressure vessel	$5 \times 10^{-7} \, y^{-1}$	$5 \times 10^{-7} \, y^{-1}$	$1 \times 10^{-5} \, y^{-1}$
Process vessel	$5 \times 10^{-6} \, y^{-1}$	$5 \times 10^{-6} \, y^{-1}$	$1 \times 10^{-4} \, y^{-1}$
Reactor vessel	$5 \times 10^{-6} \, y^{-1}$	$5 \times 10^{-6} \, y^{-1}$	$1 \times 10^{-4} \, y^{-1}$
	Atmospheric tanks		
Single containment	$5 \times 10^{-6} \, y^{-1}$	$5 \times 10^{-6} \, y^{-1}$	$1 \times 10^{-4} \, y^{-1}$
Full containment	$1 \times 10^{-8} \, y^{-1}$	$1.25 \times 10^{-8} \, y^{-1}$	$1 \times 10^{-4} \, y^{-1}$
	Pipes		
	Rupture	Leak	
Diam < 75 mm	$1 \times 10^{-6} \, m^{-1} y^{-1}$	$1 \times 10^{-6} \, m^{-1} y^{-1}$	
75 < diam < 150 mm	$3 \times 10^{-7} \, m^{-1} y^{-1}$	$2 \times 10^{-6} \, m^{-1} y^{-1}$	
Diam > 150 mm	$1 \times 10^{-7} \, m^{-1} y^{-1}$	$5 \times 10^{-7} \, m^{-1} y^{-1}$	
	Pump catastrophic failure	Pump leak	
Pumps without additional provisions	$1 \times 10^{-4} \, y^{-1}$	$5 \times 10^{-4} \, y^{-1}$	
Pumps with steel containment	$5 \times 10^{-5} \, y^{-1}$	$2.5 \times 10^{-4} \, y^{-1}$	
Canned pumps	$1 \times 10^{-5} \, y^{-1}$	$1 \times 10^{-5} \, y^{-1}$	

There are a number of other tables on heat exchangers, releases of dusts and powders in or outside warehouses, and road or tanker vehicles and for shipping. The Purple Book also contains many of the formulas and guidance for the application of those formulas found in the other colored books.

Sample outline for emergency response

Based upon our combined long experience in dealing with security and incidents, we have prepared an outline for an emergency response plan that should address many of the concerns of the regulatory agencies and should be highly useful to plant and security personnel (Table 7.5).

General note: When discussing the plant and improvements, it is imperative that only activities that are currently implemented are mentioned. Too many times, we have seen the idea that the preparation of the emergency plan, spill plan, or other document is

TABLE 7.5 Outline of emergency response plan for a typical facility

	Outline of emergency response plan for a plant
Introduction and signatures	Probably about two pages that include authorized signatures providing a brief introduction of when the plan is to be used
Plan review and reauthorization	One page showing dates where plan was reviewed and which changes were made to phone numbers and pages and other pertinent factors. Recertification of the adequacy of the plan is required in the United States for spill prevention control and countermeasure plans by a professional engineer every 3 years
RED tab section	This section should be on red border paper
	Contents should include:
Note: The purpose of this section is to centralize the notifications required in case assistance is needed and **not to have the security or guard force march down the list calling everyone**	
Calls should be made by the plant manager or his designated representative as needed or required by laws	Local emergency contact information both internal and external to the plant
Note 2: In the case of the hospital and ambulance service and the fire department, it is strongly recommended that the services should have a copy of the plan and be familiar with the plant. The hospital may need additional information about the possibility of contaminated victims coming in by ambulance, as the victims can contaminate the ambulance and/or the hospital emergency room	Emergency services
	Fire department
	Hospital and ambulance
	Plant manager
	Regional or area managers and supervisors who work in the plant
	Public relations department
	Support services such as man power for additional help with spill control and cleanup
	Plant contractor
	Regulatory and government officials including local environmental protection agency and/or state police
	Other company officials as required
	Plant maps
	General facility layout

	Location of spill control and firefighting facilities
	Inventory of firefighting equipment available
	Location of personal protective equipment and quantities
	Plant sewer system
	Size and location of tanks or flammable materials storages
	Topographic map of the facility
Emergency evacuation plans	This should be a worked-out section of multiple events for worst-case disasters and should include emergency evacuation areas and populations in the vicinity of the plant
	It should also show the reassembly areas for plant personnel in case of an evacuation, fire, or explosion
	Ideally, this section would contain contingency plans for several worst-case scenario events, including spills, explosions, vapor clouds, etc.
Regulations and regulatory compliance	This section should demonstrate compliance with appropriate environmental regulations regarding spills, releases, and incidents
Environmental contamination and mitigation information	This section should contain information on the plant wastewater treatment system, the local sewer, and the local sewerage treatment plant, including appropriate contacts for pollution control limits
Cleanup and disaster remediation and repair facilities and resources	In the event of a spill to waterways, booming and cleanup supplies might be needed. Find a contractor with supplies that are immediately available
	In the event of a fire, which contractors are good at repairing damage to the plant and rebuilding the plant
	In the event of a release to the surface or the groundwater, which companies and contractors can be counted on to provide remediation services on a cost plus basis
	Note: Especially with contract labor, it is a good idea to have a contract for services in place at agreed-upon pricing, as the multiplier for services, especially in an emergency, may be significantly higher than the normal consulting or contracting rates. The same is true for supplies but to a lesser extent

(Continued)

TABLE 7.5 (*Continued*)

	Outline of emergency response plan for a plant
Security guidance and activities during and after an incident	This should be a plan that highlights the role of the security and guard force during a plant emergency. The plan should include plant border control, who to let in, which procedures to follow, and what must be maintained at all times
Results and lessons learned from periodic tabletop and actual emergency response exercises	This section should contain summaries and recommendations from previous emergency fire drills, spill drills, and other emergency exercises. The preferred method is to present the improvement recommendations without a timetable for their implementation. Also be careful about providing detailed information about the failures from the exercises.
	Improvement projects in a plant environment can be delayed, schedules change. If past failures and planned improvements are discussed, they will provide a Regulatory Inspector with a roadmap to the problem areas, and that may effectively prepare a record of past failures which can be used against the plant in a legal or regulatory proceeding.

a way of training the new engineer or employee. As a consequence, that employee will often make comments about what should be in place to improve plant performance in a particular area. Often, those improvements are not implemented, and the "hungry" regulator who is looking for reasons to fault the plant can seize upon the "incriminating promise" statements as proof that the plant is lax in the area of protection of its employees, the surrounding population, or the environment, especially if the problem or emergency was in one of the areas where "promised improvements" were never installed.

NOTES

1 "There's Only Ethics," an address by Rushworth M. Kidder, can be found on the Institute for Global Ethics website as a free download: www.globalethics.org/resources/Theres-Only-Ethics/141/.

2 A safety culture is part of the good operation of any facility. The question these findings raise is: "What is the role of the security force, if any, in enforcement of plant-wide safety?" Traditionally, there tends to be a separation between the safety and security departments, with loose and informal information sharing. However, given the role of modern communications, potential IT breaches in the process areas, and the need to protect process facilities from sabotage, how should the security and safety departments work together to provide a safe and secure workplace?

3 NASA has a very good fault tree construction guide: http://www.hq.nasa.gov/office/codeq/risk/docs/ftacourse.pdf.

4 Much of the work on Markov chain and fuzzy fault tree analysis has been developed in the electrical engineering community and relies heavily on matrix analysis. See the following articles:

http://ieeexplore.ieee.org/xpl/login.jsp?tp=&arnumber=4126328&url=http%3A%2F%2F ieeexplore.ieee.org%2Fxpls%2Fabs_all.jsp%3Farnumber%3D4126328

Tyagi SK, Pandey D, Tyagi R. Fuzzy set theoretic approach to fault tree analysis. Int J Eng Sci Technol 2010;2(5):276–283

www.ijest-ng.com and Li Y, Huang H-Z, Zhu S-P, Liu Y, Xiao N-C. An application of fuzzy fault tree analysis to uncontained events of an aero-engine rotor. Int J Turbo Jet-Engines 2012;29(4):309–315. Available from http://www.relialab.org/Upload/files/An%20application%20of-Li.pdf. Accessed 2014 Oct 14.

5 In organization, and in execution, the Markov chain is almost indistinguishable from the fuzzy logic approach cited earlier.

6 There are two international standards dealing with the Markov approach. Both are found in the literature of the International Electrical Code, or IEC. The standards are IEC 61165 and IEC 61508. IEC 61165 provides guidance on the application of Markov techniques to dependability analysis. IEC 61508 has revitalized the need for Markov analysis due to the standard requirement to analyze different failure modes, ranging from *failed safe, detected* to *failed dangerous, undetected*. The IEC 61508 deals with the reliability of electrical subsystems. It is broken down into seven different parts which can be directly applied to security systems and subsystems. These can be directly applied to security systems and subsystems.

Markov chain analysis software includes SHARPE® and MKV. Version 3.0 (Markov Analysis Software). See http://theriac.org/DeskReference/viewDocument.php?id=95 for a description of the programs.

7 There are a number of commercial programs and books on Bayesian networks. Some of the programs require annual licenses, and some are outright purchases. The book *Risk Assessment and Decision Analysis with Bayesian Networks* by Fenton and Neil (CRC Press) is an excellent and highly recommended reading. The authors have developed a program for BN analysis called AGENA, which is supported by the www.agena.co.uk website. The program is AgenaRisk, and according to current quotations, it is licensed at about £1200 per year (2013 pricing as per their website). There is a limited version that is available if one buys the book and a more highly limited version available for free on the website.

A number of other books and programs are available for using BN. Those include WinBUGS, a free statistical program from the GNU project, and R, which is also a free software as part of the GNU project. Both are capable of completing the calculations required for BN, but the graphical interface may not be as good as that available with AgenaRisk.

The subject of Bayesian network analysis is quite complex and that the calculations are best performed by a computer, but the information obtainable about the levels of risk based on posterior analysis from marginally related events makes the analysis worthwhile. A brief sample of such an analysis is presented in the text.

8 See http://asq.org/learn-about-quality/cause-analysis-tools/overview/fishbone.html.

9 Appendix 1 Deepwater Horizon Fault Tree Investigation.http://docs.lib.noaa.gov/noaa_documents/NOAA_related_docs/oil_spills/BP_report/appendices_AA_Z/Appendix%20I.%20Deepwater%20Horizon%20Investigation%20Fault%20Trees.pdf.

10 There are a number of Internet references on preparing FMEA. Some of those include the Institute for Healthcare Improvement who has a computerized program that is free and the ASQ. It should be noted that the ASQ has a database that contains failure modes and effects failure rate data. The ASQ FMEA website also has an excellent Excel® worksheet that is more detailed than the IHI website and that is better suited to industrial process analysis.

11 Source: Iannacchione A, Varley F, Brady T. *The Application of Major Hazard Risk Assessment (MHRA) to Eliminate Multiple Fatality Occurrences in the US Minerals Industry*. Spokane (WA): National Institute for Occupational Safety and Health; 2008. Available at http://www.cdc.gov/niosh/mining/UserFiles/works/pdfs/2009-104.pdf. Accessed 2014 Oct 14.

12 At one time, the residents of Terre Haute, Indiana, had the misfortune of having a chemical plant in town that made trimethyl amine, and when the wind was in the right direction, even a small leak in the plant would cause the entire downtown to smell strongly of dead fish. The odor is persistent as well as pungent.

13 The reference documents are as follows:
    (a) 29 CFR 1910.119, Process Safety Management of Highly Hazardous Chemicals; Final Rule; February 24, 1992, Federal Register Vol. 57, No. 36, pp. 6356–6417.
    (b) OSHA Instruction CPL 2.45B, June 15, 1989, the Field Operations Manual (FOM).
    (c) OSHA Instruction STP 2.22A, CH-2, January 29, 1990, State Plan Policies and Procedures Manual.
    (d) OSHA Instruction CPL 2.94, July 22, 1991, OSHA Response to Significant Events of Potentially Catastrophic Consequence.
    (e) OSHA Instruction ADM 1-1.12B, December 29, 1989, Integrated Management Information System (IMIS) Forms Manual.
    (f) OSHA 3133, *"Process Safety Management" Guidelines for Compliance*.

14 The helpful references are found by searching the OSHA.gov website for "HAZOPS" and searching the additional links.
    https://www.osha.gov/dte/grant_materials/fy09/sh-19479-09/08_PSM_Auditing_Checklist.pdf and https://www.osha.gov/dte/grant_materials/fy08/sh-17811-08/3_psm_process_hazard_analysis2.ppt.

15 ALOHA is a part of the CAMEO suite for planning emergency response activities for chemical incidents: CAMEO has been designed by NOAA, USEPA, and other agencies for use in responding to chemical emergencies. The entire package is designed primarily for US use, but it contains a feature where one can input local mapping for planning purposes. The output from the ALOHA part of the suite allows fairly rough prediction of effects of toxic

clouds, blast radii, fireballs, and even evaluation of evaporation clouds from spilled materials. There are also related links that permit the projection of the movement of an oil or chemical slick on open water. The entire CAMEO suite can be downloaded from http://www2.epa.gov/cameo/what-cameo-software-suite.

16 As part of an attempt to evaluate the program, we decided to run a test case using information from Bhopal, India, to see if we could get results that could approximate the release. The Bhopal release of MIC was estimated by one source as almost 80 tons of MIC. The program could not model that magnitude of release. Otherwise, we found it entirely satisfactory and have used it extensively to model protective conditions for schools.

17 The difference between inherent and residual risk is that inherent risks cannot be responded to. The response can be categorized as (i) tolerating the risk, (ii) treating risk in an appropriate way to constrain the risk level to an acceptable level, (iii) transferring the risk, and (iv) terminating the activity giving rise to the risk.

18 The NIOSH Guide to Chemical Hazards can be found on the World Wide Web at the following address: http://www.cdc.gov/niosh/npg/. Additionally, the NIOSH guide is in PDF format and is also available in online format guide for emergency responders. There is also a very good guide that was mentioned earlier: http://response.restoration.noaa.gov/oil-and-chemical-spills/chemical-spills/response-tools/chemical-reactivity-worksheet.html.

19 The guidance in the Purple Book is as follows:

1. *A vessel or tank consists of the vessel (tank) wall and the welded stumps, mounting plates, and instrumentation pipes. The loss of containment (LOC) covers the failure of the tanks and vessels and the associated instrumentation pipework. The failure of pipes connected to the vessels and tanks should be considered separately.*

2. *The failure frequencies given here are default failure frequencies based on the situation that corrosion, fatigue due to vibrations, operating errors, and external impacts are excluded. A deviation of the default failure frequencies is possible in specific cases.*

   - *A lower failure frequency can be used if a tank or vessel has special provisions additional to the standard provisions, for example, according to the design code, which have an indisputable failure-reducing effect. However, the frequency at which the complete inventory is released (i.e., the sum of the frequencies of the LOCs, G.1 (instantaneous) and G.2 (continuous, 10 minutes)) should never be less than 1×10^{-7} per year.*

   - *A higher frequency should be used if standard provisions are missing or under uncommon circumstances. If external impact or operating errors cannot be excluded, an extra failure frequency of 5×10^{-6} per year should be added to LOC G.1, "instantaneous," and an extra failure frequency of 5×10^{-6} per year should be added to LOC G.2, "continuous, 10 minutes."*

   - *See the Purple Book Page 27 of 237 Section 3.3 Table 3.3 Frequencies of LOC for Stationary Vessels.*

CHAPTER 8

SECURITY SYSTEM DESIGN AND IMPLEMENTATION: PRACTICAL NOTES

SECURITY THREAT-LEVEL FACTORS

There are various factors that may influence the security of any industrial facility, as was discussed in detail in the previous chapters. These factors vary from system- or plant-related threats, to chemical–biological threats, which may directly affect internal security as a whole.

Security within the industrial plant or oil facility may also be affected by external threats, ranging from planned terrorist activities, to sabotage and individual attempts by disgruntled staff of opponents in the market. Terrorist activities are becoming more prevalent where facilities sustaining sensitive and/or expensive operations may fall victim to such attempts, especially in light of most conflicts being more asymmetric of nature (asymmetric warfare: conflicts that are nonconventional and difficult to define; the intended hostility and identity of the belligerent are not visible).

CONSIDERED FACTORS

Existence of a terrorist threat and its ability to gain access to a given facility will be influenced by the following factors and should be analyzed according to its perceived manifestation and the prevailing trends and tendencies as mapped within a specific country and certain political arenas.

The acquired assessed or demonstrated abilities or capabilities of the terrorist group must be analyzed in detail by utilizing security information analysis. It may also be advisable to approach existing sources of intelligence gathering on regional and national levels.

The intentions of the terrorist organization must be obtained or assessed. These may be found in recently demonstrated company hostile activities or intent, or stated intent to demonstrate hostility toward the company or a country. If the company represents a symbol of the country of may be classified as a national key point (NKP),

Industrial Security: Managing Security in the 21st Century, First Edition. David L. Russell and Pieter C. Arlow.

TABLE 8.1   US Department of Homeland Security color code: security threat levels

	DHS color code	
Critical	Red	Factors, 1, 2, and 5 present, maybe 3 and 4 also
High	Orange	Factors 1, 2, 3, and 4 present
Medium	Yellow	Factors 1, 2, and 4 present
Low	Green	Factors 1 and 2 present, factor 4 maybe present
Negligible		Factors 1 and/or 2 present

the threat potential will be far higher than expected. The following are critical considerations: What is the history of the group? Have they conducted terrorist activities in the past? What is the current credible information on activity indicative of preparations which indicate attack is imminent? The US Department of Homeland Security (DHS) uses the following color code to assist during the analysis of threat levels (Table 8.1).

Terrorist or other hostile activities may be attempted to destroy a facility or to affect its specific output of a specific chemical or product. In most cases, these activities are conducted by using any form of explosives, varying from manufactured munitions to improvised explosive devices. Vehicle bombs are best known for these attempts, due to its mobility and opportunity to enter a facility through the standoff zone and will be discussed later.

Vehicle bombs

Vehicle blasts can develop very high pressures, but the pressures decrease rapidly with distance. This was clear during the highly destructive nature of the explosion at the Khobar Towers in Riyadh—1996. The terrorists were reported to have smuggled explosives into Saudi Arabia from Lebanon. In Saudi Arabia, they purchased a large gas tanker truck and converted it into a bomb. Al-Mughassil, Al-Houri, Al-Sayegh, Al-Qassab, and the unidentified Lebanese man bought a tanker truck in early June 1996.

Over a 2-week period, they converted it into a truck bomb. The group now had about 5000 lb of advanced, high-grade plastic explosives, enough to produce a shaped charge that detonated with the force of at least 20,000 lb of TNT, according to a later assessment of the Defense Special Weapons Agency. The power of the blast was magnified in several ways. The truck itself shaped the charge by directing the blast toward the building. Moreover, the relatively high clearance between the truck and the ground gave it the more lethal characteristics of an airburst. It was originally estimated by US authorities to have contained 3000–5000 lb of explosives.

Later the General Downing report on the incident suggested that the explosion contained the equivalent of 20,000–30,000 lb of TNT. The terrorists prepared for the attack by hiding large amounts of explosive materials and timing devices in paint cans and 50-kg bags, underground in Qatif near Khobar. The bomb was a mixture of gasoline and explosive powder placed in the tank of a tanker truck (Figs. 8.1 and 8.2).

Figure 8.1 One view of Khobar Towers bombing (Riyadh, Saudi Arabia) in 1996. From http://web.ornl.gov/~5pe/p022/khobartowers.jpg.

Figure 8.2 Damage at Khobar Towers, note size and depth of bomb crater. Defense department photo: http://airforcelive.dodlive.mil/2013/06/everyone-has-a-story-the-grocery-bagger/.

In order to prevent such an attack, the barriers on the perimeters must prevent penetration of the vehicle beyond the standoff zone. There are three levels of protection based on building damage:

1. Low: Buildings destroyed
2. Medium: Buildings damaged but re-usable
3. High: Only superficial damage (buildings may be designed to be specially blast resistant)

There are two very important measures to be instituted of improved to ensure that barriers and perimeters are impenetrable:

1. Reinforced vehicle barriers
2. Planning and maintenance of the standoff distance

It is however important to first understand the levels of threat and types of weapons that may be used to attack a facility, before the importance of vehicle barriers and the enforcement of the proclaimed standoff zone and distances could be understood best.

Standoff weapons

Standoff weapons include machine guns, artillery, heavy caliber guns, and mortars. They cannot be detected in advance. The best protection is prevention of line of sight from exterior vantage points. Screen with trees, walls, or fencing. Even wooden fences can be used for pre-detonation devices.

If mortars are a concern, strengthen roof surfaces to withstand blast and add a layer of protection to reduce line of sight. The facility may require added internal reinforced concrete walls to serve as sacrificial walls for blast resistance. Masonry is generally resistant to all but armor piercing rounds. For military weapons, thicknesses must be doubled to 18–20 cm of brick or 18 cm of reinforced concrete.

Minimum standoff distances

The minimum standoff distance is 50 ft (20 m) or more depending on the type and amount of explosives anticipated. Use tools such as **Aloha** or **Archie** (disaster management tools or the formulas presented in earlier chapters) to calculate blast damage based on the size of the vehicle and anticipated weight, and adjust accordingly. Walls may tend to magnify blast and can create missiles if blast is next to wall (Figs. 8.3 and 8.4).

For nonexclusive standoff zones, an additional layer of distance (protection) is required (see Fig. 8.5).

The best protection is achieved thorough search and not allowing any vehicles inside the standoff zone, unless cleared by security after searching.

Fencing is not a barrier. Most fencing will delay people less than 10 seconds, 4 seconds to climb and 10 to cut. A bomb inside a building (mail room) is much more hazardous than a bomb outside because the force is not dissipated. Keep the mail room and delivery points separate and rather of light construction or revetment grade

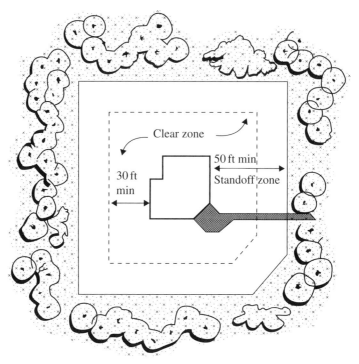

Figure 8.3 Minimum standoff zone. Note distance is a minimum depending on type of weapon attack anticipated. US Army Field Manual on Physical Security: FM-19.30 *Physical Security*.

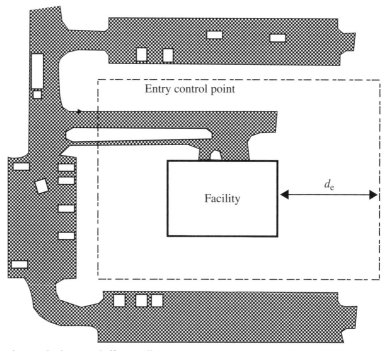

d_e = exclusive standoff zone distance

Figure 8.4 Standoff zone for medium-to-large facilities. US Army Field Manual on Physical Security: FM-19.30 *Physical Security*.

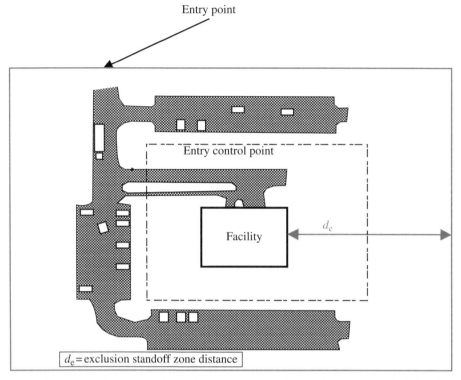

Figure 8.5   Exclusion zone for larger facilities. US Army Field Manual on Physical Security: FM-19.30 *Physical Security.*

(revetment grade is designed to withstand and dissipate a blast). Receiving areas should be kept away from other areas and designed to prevent blast damage. Utility openings, including drainage ditches, and sewer openings of greater than 20 cm diameter should be protected against intrusion or insertion of a weapon. Seal manholes and close gaps for drainage swales. Barrier fences should be at least 2.1 m tall, with barbed wire or concertina wire on the top. Maintain a clear zone around the perimeter of the exclusion zone.

SECURITY SYSTEM DESIGN

Security design is probably the most important element for the future existence and profitability of any industry or company. The security architecture is therefore dependent on detailed consideration, planning, and implementation of various facilities and infrastructure upgrades to ensure a proactive and technologically relevant security environment. Aspects to be considered, as well as discussed further in this chapter, are as follows:

- Perimeter barriers
- Vehicle control barriers

- Entry roadways and control stations
- Reinforcement of buildings and infrastructure
- Security lighting
- Electronic security systems
- Exterior sensors
- Access control
- Employee screening
- Visitor identification and control
- Personnel and packages control (deliveries and contracting)
- Lock and key control
- Security and guard forces
- Cargo security
- Port security (where applicable)

Perimeter barriers

Perimeter barriers are fixed around the perimeter of the standoff zone and include the following types which may vary according to the threat potential and level of fortification required:

- Chain link fence
- Hedges
- Curbs at least 8″ (20 cm) high
- Jersey barriers (Fig. 8.6)
- Cable reinforced bollards
- Spacing 1.2 m or less

Active vehicle barriers are much more expensive, but with proper design can stop vehicles of 7000 kg at speeds of up to 85 km/h. Much heavier construction is required which further ensures anchorage in the ground. Active vehicle systems include bollards, cable beams, sliding gates and drum, and so on.

Active vehicle barriers

Active vehicle barriers are installed and constructed at all main entry and exit points to facilitate controlled movement. Although movement control is exercised from these locations, active vehicle barriers must be able to prevent forced entry in the best possible way. This aspect is again dependent on the level of the anticipated threat and the analyzed methods of forced entry expected. In Figure 8.7, different active vehicle barriers are illustrated, varying from cable beam barriers and retractable bollards, to drum-type barriers.

Figure 8.6 One type of perimeter barrier. Photo of a New Jersey Barrier: www.safety.fhwa. dot.gov.

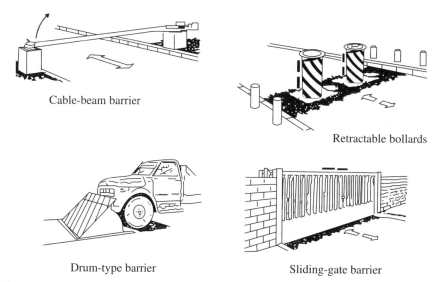

Cable-beam barrier

Retractable bollards

Drum-type barrier

Sliding-gate barrier

Figure 8.7 Active vehicle barriers. US Army Field Manual on Physical Security: FM-19.30 *Physical Security*.

Entry roadways

Speed control is important. Establish an entrance lane that provides a serpentine path and that limits vehicle speeds to 15 km/h. Use Jersey or other types of barriers to slow vehicles. Establish barriers both inside (after) and outside (approaching) the perimeter barrier. The sides of the entry roadway should have high curbs to prevent vehicles from leaving the roadway.

Entry control stations

Entry control stations should be located at main entry points where guards and control staff are present. A holding area should be established for unauthorized vehicles and the turnaround for other vehicles prior to inspection. Vehicles passing through the entry control stations should display a vehicle sticker or temporary visitor card. Entry control stations should be manned 24 hours and should be equipped with quality interior and exterior lighting. Exterior motion detectors have to be installed to enable threat detection or vehicle movement from a distance. Entry control stations should further be reinforced for it to be bullet or blast proof as the threat may indicate. First-class communications systems must be installed and made available with secure interfacing between telecom and radio facilities. Sufficient technology must be considered to assist signaling potential threat indicators. Signs clearly signaling all control requirements and law enforcement policies, where applicable, must be displayed (preferably in most local languages possible) at least 30 m from control and entry stations.

It is important to consider that the master command, communications, and control center should not be located at the main entrance unless the building is blast proof. In every event, a backup location away from the front gate should be established, and all communications should be routed to this station in parallel so that it has the same information as the main station in the event that the main station is disabled by blast, attack, or incident. All sensors should be routed to this backup station as well, but *not through the primary station*.

Reinforcement of buildings and infrastructure

Blast forces will be substantially horizontal, but will require reinforcement of floors and walls to withstand blast pressure. Most buildings nowadays are designed to withstand 2.3 kPa; explosion pressures can be significantly higher. The use of reinforced masonry or concrete is added to newer buildings to absorb or counter blast pressures.

Windows

Windows present a special problem. Flying glass accounts for 85% of injuries. The following preventative and reinforcing methods may be considered and applied according to the nature of the anticipated threat potential:

- Use fragment retentive films.
- Install blast curtains.
- Reinforce window frames or replace it with blast resistant designs.
- Design narrow windows, preferably on top of occupied space to reduce glass hazards and the likelihood of thrown or fired projectiles.

Security system lighting

Security lighting is probably one of the most important elements of security and will enhance all other technologies and planned efforts to secure any environment. Without quality and sustainable lighting in all strategic locations, security services are rendered blind. In order to plan and install a guaranteed security lighting system, the following critical considerations are applicable:

Considerations

- Cost of replacement, cleaning, and maintaining fixtures including bucket trucks for high lights
- Manual override provisions during blackouts—requires separate power sources
- Local weather and its impact
- Electrical requirements, voltage fluctuations, grounding requirements, and rapid bulb replacements
- Use of lighting for CCTV support
- Exclusion areas and critical areas (high risk)
- Protective lights—redundancy and independence so multiple failures do not occur
- Requirements of adjacent properties
- Restart time (after power failures)
- Color accuracy and illumination levels
- Parking and control areas and guard and fence areas

Lighting system design

The value of a well-planned and sustainable lighting system design is that it will assist in discouraging intrusion by making detection likely and that it will enhance all re-active efforts by the guard force and other observers.

Boundary areas must be lit so that guard paths are darker, in order to present glare in the eyes of attackers. High brightness and contrast between intruders and background is required. Though it may sound ridiculous, it is crucial to keep buildings brightly painted and clean to assist in providing sufficient silhouetting of the intruder. Standby or emergency lighting should duplicate existing patterns. Illuminating both dock areas and approaches is required. Docks should have at least 10.74 and $5.37 \, \text{lx/m}^2$ for water at least 30 m out from the pier (port/plant security).

ELECTRONIC SECURITY SYSTEMS DESIGN

Electronic Security System (ESS) is the placement and implementation of electronic systems to serve as early warning to unauthorized intrusion or other planned attacks. This may include closed-circuit television (CCTV), security lights, various forms of sensors, alarms systems, and a well-trained and adequate guard force for response. ESS should be reliable, accurate, and updated according to most recent technologies.

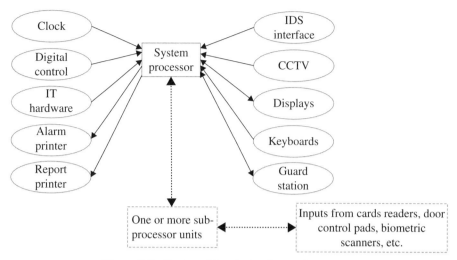

Figure 8.8   Design of the electronic security system.

ESS must delay the intruder from achieving their objectives until response arrives. A well-designed system minimizes the possibility of covert intrusion. All sensor systems have nuisance alarms and physical design constraints for detection. ***Respect those constraints!*** Manufacturers do not provide information on nuisance alarms: they occur from environmental conditions (wind, birds, etc., and from electrical faults). If alarm system sensors are delayed, it increases the area of search.

The speed in which detection is achieved is important. If a fence is scaled or cut in 10 seconds and man runs 6 m/s, a 2-minute delay could result in an area search of over 80 ha (200 acres). CCTV cameras must be independently illuminated. A scanning system should be installed for more than 10 cameras. Most exterior intrusion sensors are exposed to much more rugged environmental conditions and generally do not detect movements above 2.5 m even on fences. Buried sensors generally are not able to detect movements more than a meter from the ground surface. Interior Electronic Security System (ESS) sensors are generally less costly than external sensors. For entrances, windows, and so on, there has to be an access mode where the alarms are shut off for normal access and a secure mode where they are activated. The secure mode should never be locally controlled and access mode must not de-energize the alarm. Duress and tamper switches must never be put into access mode. Each type of sensor has its limits. Fog, rain, and dust limit infrared (IR) capabilities—it might therefore be advisable to consider thermal imaging to supplement IR. Wind may cause fence-mounted sensors to give false alarms. Vegetation can cause many false alarms and conceal intruders. Line of sight is extremely important for detection and confirmation (Fig. 8.8).

Alarm configurations and design

Alarm configurations for small systems may provide individual alarms for specific areas or a general alarm depending on configuration. An ideal system will provide specific area notifications to increase probability of detection and minimize false positives. Computer-assisted systems may use multiple computer processors and automatic reset as well as entry/card acceptance for certain functions. All alarm systems should be connected through redundant data transmission links to prevent local loss of signal from inactivating regional and zone alarms. Alarms should be logged, preferably by printer. There are five possible alarm levels:

1. Duress or life-threatening emergency alarms
2. Intrusion detection
3. Electronic entry control
4. Tamper signals
5. CCTV and equipment malfunction alarms

Exterior sensor types   Exterior sensors are quite straight forward and differ from facility to another facility, again based on the threat levels and the type of intrusion/ violation that could possibly be expected. All external sensors have the mutual objectives to provide early warning and during extreme measures to terminate the attack/ intrusion. External sensors may consist of the following:

- Fence sensors
- Strain-sensitive cables
- Taut wire sensors
- Fiber optic strain
- Electrical fields
- Capacitance proximity sensors
- Buried line sensors
- Line-of-sight sensors
- Microwave sensors
- IR sensors (active and passive)
- Video motion sensors

Access control

Access control is the primary point for the enforcement of security and is probably the most vulnerable area, providing entry to the processes of the industry or complex. The main focus of effort by security staff, the guard force, and detecting technology should be directed here. History has taught that most unauthorized entry, especially vehicle improvised explosive devices proceed through access control points. In many case studies performed, this is usually supported by staff within the complex, either being supportive to the intruder or being forced to participate. There are three main types of access control points that should be established to ensure controlled entry.
Three types of areas are as follows:

1. *Controlled area*—The area surrounding an exclusion zone entry. This area is controlled, but all movement remains unrestricted.

2. *Limited area*—An area surrounding a sensitive security interest. Escorts may be required.

3. *Restricted area*—This is the area where the security interest is located. It includes control rooms and guard facilities. Clear restriction and warnings signs should be posted outside each area.

Employee screening

All employees must be screened to eliminate potential threats. Before hiring any personnel, the following aspect should be checked and verified:

- State, local, or national police
- Former employees
- Public records
- Credit agencies
- Schools at all levels
- References not furnished by the applicant

Medical screening may be necessary to establish the mental and physical condition of the candidate. Family medical history may also be appropriate for severe medical stress or sickness.

Identification cards may be adequate for low security areas. Badges with personal details are required for areas with over 30 employees/shift personnel. Personal recognition systems (uniforms or color coding) depend on guard force protocols. Multiple badges and cards/color coding may be required for varying levels of security entry. Card or badge specifications should include designated areas where cards/badges are required. Description of the badge in use and authorization limitation of the bearer must be indicated and verified to the employee and control point. This must be presented when entering or leaving each area at all times. The disposition of cards upon termination of employee, or other causes, must also be clarified.

Visitor identification and control

Visitor identification is a critical part of access control. Due to the fact that any industry is dependent on contracted services, deliveries, shipment of cargo, and other consignments from the facility by other industries, hostile or unauthorized entry occur through this aspect. Visitor identification and the control of all cargo/items entering or leaving the facility will limit the vulnerability against any form of unauthorized/hostile intrusion. The following are the most important aspects to consider and check during the authorization of visitors to the complex/facility:

Written policies and procedures establishing visitor control

- Prearranged approvals for admission must be cleared.
- Escorts must accompany all visitors in the limited and restricted areas at all times, especially in the event of foreign nationals.

- Visitor classifications must be qualified and clearly defined.
  - Suppliers, customers, inspectors, vendors, and regulars
  - Visitors for educational purposes
  - Visiting groups of foreign nationals and guided tours, and so on
- The following is applicable to the reception of all visitors:
  - Authority must first be established whether plant personnel may receive the intended visitor.
  - Positive identification (ID) documents of the person receiving visitation must be verified (permit or credentials from employer).
  - Cards/badges must be used at all times, where applicable.
- Cleaning teams should also be screened and clearly identifiable by plant security.
- Supervisory coordination based on work hours and restrictions must be on the security schedule of the specific shift.
- Procedures for admission must be uniform and enforced.
- Limit entry/exit control points within the facility.
- Educate guard force and employees to work together on all protocols.
- Single file admission with verification must be in place.

Packages, personnel, and vehicle control

A package checking system must be enforced prior to entering all restricted areas. Inspect all outgoing packages for authorization (cuts down on pilferage). If 100% package control is not possible, use frequent random checks and inspections. Personal vehicles and packages, tool boxes, and so on, need to be inspected during entry and exit. Visitor's vehicles must be clearly marked. Truck and rail movements in and out must be inspected. Truck and rail gates must be locked. Shipment must be sealed and seals inspected upon entry. Incoming trucks and rail cars must be logged in. The following details must be logged:

- Driver's name, license, load description, and time of entry and departure.
- Check operator's license.
- Escort when necessary.
- Verify seals unbroken and unhampered with.

Lock and key systems

Key locks are only good for low security systems and offices. Dead-bolt locks and mortise locks are only slightly better than straight key locks. Drop-bolt locks are better than dead bolt. Combination locks need to be backed up by other locking devices when area is unoccupied. Padlocks are mostly low security devices, except high security padlocks that have hardened parts. *ALL LOCKS ARE DELAYING DEVICES AND IS NOT A POSITIVE BAR TO ENTRY OR FORCED ENTRY BECAUSE THEY CAN BE DEFEATED!*

Security forces

There is a vast difference between security staff recruited from a local home grown origin and that of a contracted nature. The type of security required versus the potential threat to the facility will determine the type of guard force required to protect the security interest. Factors like the origin of candidates, qualifications, and cost obviously have to be considered, but the perceived threat potential will ultimately determine the type of security forces needed. You should answer the following questions: Do you want a local rent-a-cop or a professional. There is a difference in cost and level of involvement.

It is further important to determine the levels of authority and jurisdiction. What special powers or authority is required to effect arrests? What jurisdiction will the guard force have in lieu of existing policing and/or defense forces jurisdiction? What other armed force response are available to contribute to the capacity of a guard force? Consider liabilities for accidental deaths. Relations with local police and military are important. Consider force organization and response when co-coordinating roving patrols. Who responds, with what, where and how many?

Standard operating procedures (SOPs) have to be drafted and implemented to guide and control all security force activities within the facility. It must be designed to clearly stipulate procedures, responsibilities, accountabilities, and roles, especially in the event of emergencies, attacks, and other unauthorized activities. The security forces must be controlled from a centralized command and control location, in some cases referred to as an Operations Control Center or Headquarters, and all staging and forming-up areas must be known and rehearsed, as part of contingency planning.

Security personnel must have provisions for shelters, relief, and breaks (at least every 2 hours). Security Personnel may only be utilized for security, not firefighting (unless in an emergency)—but cross training for use when off duty is permitted. Strict instructions and posting assignments must be issued, as well as for actions required during emergencies elsewhere in the plant.

Training must be supported with regular evaluations, testing of skills, and rehearsing of drills. Security forces may require uniforms, specialized vehicles and equipment, dedicated communications infrastructure and radio equipment, traffic control equipment, sirens, flashlights, weather gear, and so on. Training should include the following:

- Areas of responsibility
- First aid and fire control equipment operation
- Common forms of sabotage and espionage
- Locations of hazardous equipment and material within the plant
- Weapons where required and legally proportional and appropriate

Cargo security

Harbors, ports, and terminals are highly susceptible to security breaches because of high levels of foreign (non-plant) workers and movement of goods. Security needs to monitor the area and establish a perimeter and classifications for various personnel.

Patrols should be combined randomly and regularly. Specialty (high-value) areas should be clearly designated and considerations include the following:

- Type and value of cargos stored
- Vulnerability of cargo to land threat
- Likelihood of diversion, sabotage, theft, and so on
- Location and nature of ports and cargos
- Degree of entry and exit controls

Port security systems

Keep cargo secured while being transferred. Establish security perimeter and access control points. Erect field expedient barriers and limit personnel access to those required. Provide a separate holding area whilst truck cargo is inspected and sampled where required. Inspect inbound and outbound containers. Verify records, seals, and documentation. Respond to various threat levels with appropriate security measures.

REVIEW AND ASSESSMENT OF ENGINEERING DESIGN AND IMPLEMENTATION

Auditing and evaluation

Continuous auditing and evaluation of security systems is critical to ensure that the most appropriate and updated systems design is maintained at all times. Due to the fact that the threat scenarios continuously change in any facility, auditing and evaluation must be formal processes allocated to an accountable team within the security environment. A risk assessment team must be appointed and should consist of the following staff:

- *Risk Assessment Manager.* The risk assessment manager is accountable for the continuous threat analysis and risk assessment within the facility. He/she will direct all activities-related threat analysis, security systems design, and the implementation of required upgrades and rectifications.
- *System Administrator.* The system administrator keeps record of information, requirements, and future systems design, on behalf of the risk assessment manager. All administration has to be logged electronically, preferably using a system that keeps a paper trail of all findings and recommendations by the risk assessment team.
- *Technical Reviewer.* The technical reviewer is qualified in the continuous testing and evaluation of all security-related systems within the facility.
- *System Business Advisor.* The system business advisor gives recommendations toward the financial situation within the security environment and deals with the allocation of funds and the budget.
- *System Technical Advisor.* The system technical advisor is responsible for analyzing all information during the auditing and evaluation of the security system design and to formalize recommendations for future required adjustments.

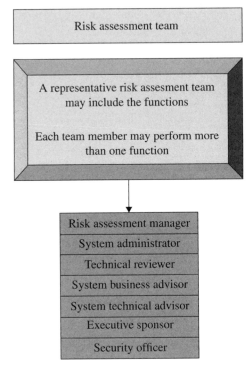

Figure 8.9 Security staff and committees to be trained and instituted as a risk assessment team.

- *Executive Sponsor.* The executive sponsor could range from the budget holder within the company, to a body of trustees or even external sponsors who may have interest in the capacity of the company and/or the necessity for a secure environment.
- *Security Officer.* The chief security officer has to be co-opted onto the team or committee due to his responsibility for the implementation and management of security within the complex.

Risk assessment team

Figure 8.9 illustrates the preferred groups and individuals which should be incorporated into the Risk Assessment Team. The figure relates to Electronic Security System Design. The figure is an graphical representation of the Electronic Security System Design Elements.

Security management Security management is an integral part of management as a whole. The executive staff of the facility remains accountable for security, even though qualified security staff is employed and appointed to fulfill different responsibilities within the security environment. Figure 8.10 indicates the relationship and channels of liaison from a management perspective.

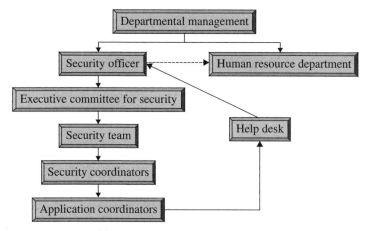

Figure 8.10 Security management.

Blank sheet approach to auditing and evaluation

The blank sheet approach to auditing and evaluation is the most effective model to implement in order to maintain a secure sequence for the identification of challenges within the security environment and the continuous rectifications and implementation of required upgrades.

The blank sheet approach provides a cycle of activities, which will continuously start and end, to ensure a live and frequent analysis of the security systems design, as follows:

- *Identify*. Identify the needs of the system, and then identify the related risks.
- *Understand and agree*. Ensure an understanding of what has to be implemented. Agree on what has been found as risks and what needs to be implemented.
- *Solutions*. Find solutions for the risks identified and agreed upon.
- *Manage*. Provide the necessary advice and tools (if need be) to manage the risk.
- *Evaluate and report*. Once the risk is accessed, evaluate the management thereof and provide a report.
- *Audit*. Provide periodical audits as required. If new risks are identified, the process repeats itself (Fig. 8.11).

Business approach to auditing and evaluation

The blank sheet approach to auditing and evaluation, as discussed earlier, is more informal and provides a logic cycle of assessment and rectification. The business approach is a more formal system related to the management activities and processes within the facility. We will examine this again later in this chapter.

Figure 8.11 Blank sheet approach to auditing and evaluation from inception through implementation. A continuous and cyclic process.

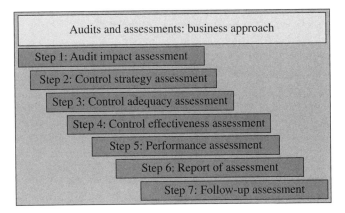

This approach deals with already identified risks and security issues, as well as identifying and gaps. It is based on the Institution of Internal Auditors Business Approach to Auditing

Figure 8.12 Business approach to auditing and assessments.

The business approach to audits and assessment is a list of steps to be followed from the audit impact assessment down to the assessment report and follow-up assessment (see Fig. 8.12 for the business approach to audits and assessments).

Benchmarking

Benchmarking is a continuous ongoing long-term process. It is a systematic, structural, formal, analytical, and organized process for evaluating, understanding, assessing, measuring, and comparing business practices, products, services, work processes, operations and functions of organizations, companies, and institutions that are recognized, acknowledged, and identified as best-in-class, world-class, and representing best practices for the purpose of organizational comparison, organizational improvement, meeting or surpassing industry best practices, developing products/process objectives, and establishing priorities, targets, and goals. (Source: Van der Zee HTM. *Measuring the Value of Information Technology*. Hershey (PA): IRM Press; 2002: p. 144.)

How to evaluate a physical security system?

A security system is more than the sum of its parts. The components of the system are just the basics. The system must address more than just fence line intrusions. There is a strong personnel component in any security system. It must be flexible and secure at the same time. The security system must be capable of considering multiple elements including natural disasters (typhoons, sand storms, Tsunamis, earthquakes, etc.), industrial accidents, including sabotage and arson, criminal acts (arson, theft, etc.), terrorism, and other possible scenarios.

A security system must consider assets, exposure, loss, and loss prevention within the framework of limited costs and personnel interactions and liabilities. A totally secure system is an empty tank in an abandoned plant. Activity incurs risk!

Security systems audits

A good audit is a thorough examination of all parts of a system and tests the system for response to activities. A good audit is more than a paper trail, but the paper trail is important. A good physical security system includes interviews and thorough physical examination of the mission and the system being evaluated. It is both active and passive and requires a team to evaluate.

A good security system must plan for the unthinkable and undesirable, and must be able to integrate internal and external organizations which function in its support such as the following:

- *Hospitals*. Patient decontamination, transport, and equipment available.
- *Fire*. Type of response available—is it suited to the plant needs?
- Police, security, intelligence, crime scene investigation, and capturing of terrorists.
- *Military*. Is the facility of critical interest? Does the military need to be involved in the response? Bomb disposal?

Define types of risk to be assessed and types of effects from incidents. Define the probability of occurrence. Prioritize the loss potential, interview personnel, review files, collect and analyze data, and compile a detailed report.

Conduct a preliminary data gathering effort. Obtain the mission statement and directives for the security function. It should be part of the overall company mission statement. It should also have a specific function and responsibilities. Interview long-term employees regarding incidents and activities; include management personnel. Oral history and written records is of tremendous value. Include retired employees where possible. Interview and record information as part of the database must be accessed. Observe and inspect security measures. Conduct a physical inspection and finalize the security audit.

Gather assets, exposure, and loss data from the corporate risk manager and controller's offices.

Fixed assets	$_____
Owned, leased assets	$_____
(Less) Facility losses	–$_____
= Total tangible assets	+$_____
Total intangible assets	$_____

This may include various categories of exposure and collateral and contributory losses and liabilities, for example, losses from business interruption, replacements, cleanup and decontamination, disposal, and other sources.

Types of losses to be considered may also include the following:

- Crime
- Cargo pilfering and damage
- Emergencies and disasters (earthquake, etc.)
- Damage
- Environmental controls and regulatory fines
- Liability of officers
- Business interruption
- Errors and omissions (negligence)
- Professional liability (third party on your property)
- Product liability (not usually considered except by risk manager and lawyers)
- Personnel and kidnapping

What to review?

The following aspects must be taken into consideration during the review and assessment of the security system within the facility:

Policy and programs

- Policy directives as written document
- Clear assignment of responsibilities by position
- Designated (position) individuals
- Is the top manager available to the security director?
- Published, clear regulations, and directives

Written disciplinary procedures

- Written description of offenses
- Includes written description of penalties and offenses

- Uniform enforcement of policies (Trojan Powder policies)[2]
- Actions must be recorded
- Actions must be reviewed by upper management
- Sign off by upper management is required

Published, clear regulations and directives
- Especially important for multiple facilities
- Must be available to all employees

Operations
- Full-time security supervisor or only a percentage of time is spent on security matters.
- Chain of command direct to plant manager.
- Number of personnel (adequate?/shift?).
- Other duties of security personnel. (Do they have a priority for security or something else?)
- Training: What training, documentation and type.
- Written incident reports with full documentation.
- Follow-up (after action) reports and records.
- Policy on criminal prosecution of violators?
- Background checks on security personnel.
- Guard force adequacy.

Guard force adequacy
- Are tours electronically tracked?
- Do guards make written reports?
  - Secretarial assistance.
  - Permanent records.
- Tour frequency and pattern (varied pattern?).
- Number and location of reporting stations?
  - How do you make sure the guard is doing his job?
  - Key stations?
  - Who gets reports when the guard is on rounds?
  - Is there a backup system?
- Written reports on each shift? Reports reviewed?
- Guard training (especially if weapons authorized).
- Condition of uniforms.
- Numbers of guards for each shift balance against threat levels.
- In house guard force or contracted guard force?
- If contracted, does the plant security director select spot checks?

- What qualifications and what criteria are required?
- Are written contract and orders in place?
- Are weapons carried?
- Who inspects and maintains weapons?
- Who furnishes weapons?
- What type of weapons?
- Weapons training for security personnel.
- The level of the security threat and outside response must be balanced.
- If weapons are carried, policy must be written and approved—think about the guard who has a bad day and may be irritable?
- Posting about weapons warnings—is it clear and prominent?
- Published security procedures and how are they distributed?
- How frequently are procedures revised?
- Do guards and security forces conduct drills?
- Does security supervisor maintain contact with guard force?
- Local police and military (if required)?
- Is security supervisor aware of local community discontent or disorder or criminal activity?
- Entry control and movement.
- What type of plant barrier is there to prevent intrusions and is it continuous?
- Are non-barrier areas illuminated at night and observed?
- What are the non-barrier areas?

Fencing

- Three meters high?
- Five-centimeter mesh or difficult to climb?
- What is wire gauge (thickness)?
- Fastened to fence posts securely?
- Barbed wire on top?
- Fence posts set securely (concrete or depth)?
- Gates same height and construction? When open (only during use)? Deliberately locked after use? Observed and guarded when open? How guarded?
- At least 3 m clear zones each side of fence?
- Is the fence on top or at least 7 m from bottom of embankments?

Walls

- At least 3 m tall?
- Topped with razor wire or barbed wire?
- All doors equipped with alarms?
- Means of observation?

Windows

- Permanently closed?
- Accessible for removal of property?
- Can they be used for entry or exit?
- Protected by bars? Alarmed?

Perimeter doors

- Guarded?
- Alarmed? Specify type.
- Security controlled?
- Impact resistant?
- Hinges and locks tamper proof and non-removable?
- Do outside door locks have a keyway? (NO!)
- Are the doors dead bolted?

Lighting

- Is the entire perimeter lighted? On both sides of the fence?
- Is the illumination sufficient to enable the detection of human movement at 100 m?
- Are the lights checked daily before it becomes dark?
- Are burned-out lights replaced immediately?
- Is the power supply for the lights tamper-proof and readily available to the guard force?
- Are switches and controls readily available but tamper proof, weather proof, and inaccessible from outside, with a centrally located master switch?
- Is there good illumination for guard routes inside the fence?
- Are the materials receiving and shipping area sufficiently lighted?
- If there are docks or bodies of water, is the illumination sufficient to enable detection of movement?
- Is there an auxiliary power source for lighting?
- Recommended lighting levels:

	Illumination	
Condition	(ftcd)	(lx)
Full daylight	1000	10,752.7
Overcast day	100	1,075.3
Very dark day	10	107.53
Twilight	1	10.75
Deep twilight	0.1	1.08
Full moon	0.01	0.108
Quarter moon	0.001	0.0108
Starlight	0.0001	0.0011
Overcast night	0.00001	0.0001

Activity	Illumination (lx, lumen/m^2)
Warehouses, homes, theaters, archives	150
Easy office work, classes	250
Normal office work, PC work, study library, groceries, show rooms, laboratories	500
Supermarkets, mechanical workshops, office landscapes	750
Normal drawing work, detailed mechanical workshops, operation theatres	1000
Detailed drawing work, very detailed mechanical works	1500–2000

Locks and keys

- Are all locks and keys under control of security supervisor?
- Who has authority for changes in locks and keys?
- Is there a written policy for key and lock issuance? Is it approved?
- Do nonemployees have keys?
- Is the issuance of keys documented, reviewed, and approved by appropriate management?
- When employees leave, are they obligated to turn in keys?
- Are master keys unmarked?
- Are spare keys kept under double lock?
- Locks on perimeter doors changed annually?
- Manufacturer's number on locks obliterated and changed to plant number?
- When gates and doors are opened, are locks relocked to prevent substitution?
- Regular checks for tampering?
- Do door lock bolts extend at least 1.5 cm into jamb to prevent tampering?
- Is the bolt covered by a tamper proof plate?
- Are combination locks regularly changed? How often? When employees leave?
- Are combinations memorized (must not be written)?
- Are combinations in any specific sequence?
- Are combinations disclosed on the basis of operational necessity rather than convenience?

Safes   Are safes substantial, fireproof, rated, lighted (24 hours), and covered by motion detectors?

Fire alarms

- Water flow? Water pressure? Valve open or closed?
- Combustion detection, heat or smoke sensing. How is it monitored? Continuously monitored? Directly connected to fire or police? Regularly tested and documented?
- What else do the alarms do? (Shut off power, A/C, lights, etc.)

Intrusion alarms

- Plant perimeter?
- Type of alarm? (motion detection, heat sensing, and other)
- High-value storage areas and type of alarm?
- Types of sensors in each area (list and check for adequacy)?
- Are alarms reported to a central station?
- Are alarms tested regularly and results documented?

Communications

- Separate and multiple methods of communication for guards—telephone, radio, and so on.
- If radio shared does security have an override?
- How is request for help given to outside parties? Can it be interrupted?
- Plant-wide signal for an emergency condition? Specify the condition and the signal.

Property control

- Procedures must be written, must use specific forms, serially numbered, auditable, authorized (by higher authority).
- All property transfer actions monitored at exit?
- All exits controlled?
- Control points and inspection between parking and work area on incoming packages?
- Spot checks on trucks?
- Company tools clearly marked? Issued with receipts?
- Lost equipment and tools reported? Follow-up investigation? Written record? Monitored?

Shipping and receiving areas

- Guarded and surveilled or within protected areas?
- Inspected regularly and spot checks?
- Written policy on outside drivers in plant?
- Checks on all incoming and outgoing vehicles, including tailgate checks?
- Are storage areas monitored and all withdrawal receipts checked and audited?

Scrap and salvage

- Procedures for collection and disposal must be written. Bids must be sealed, and independently reviewed. Scrap may never be given to employees without management approval.
- Wastes must be stored in a locked area, spot checked for saleable materials and high grading.

- Wastes may be removed only:
  - With signed authorizations, checked, and compared with receipts.
  - Supervised by security and spot checked.
  - Auditable on the basis of transactions.

Personnel

- Employees must apply on approved forms.
- Candidates must be interviewed and should include the following:
  - Previous employers and dates
  - Position and duties (watch out for inflation)
  - Salary and quality of performance (supervisors' names)
  - Education (verified)
  - Criminal record (verified)
  - Reputation (investigated)
  - Medical records:
    - Illnesses
    - Handicaps
    - Work injuries
    - Occupational illnesses
- All positions, especially financial, must be investigated and background checks performed.

Review the emergency and disaster plan for completeness and contingencies   The Emergency and Disaster Plans are integral to the operation of the facility. The plans should be reviewed and updated periodically to insure that the actions and contacts and equipment required for emergency response are all in good repair and usable in an emergency. It is also necessary to drill on these plans for a number of different contingencies.

Implementation of risk assessment

Risk assessment process flow   The risk assessment process flow is depicted in Figure 8.13 in three phases, as follows:

The risk assessment project   The phases for the risk assessment flow are further followed to outline the different timelines or sequencing of the risk assessment project as follows (Fig. 8.14).

Severity of impact and risk levels   During the risk assessment and auditing process, the severity of the perceived or expected impact of the risk identified and the levels of intensity thereof must be compared according to the scales indicated in Table 8.2.

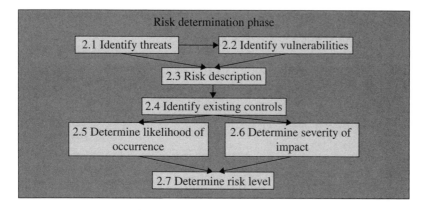

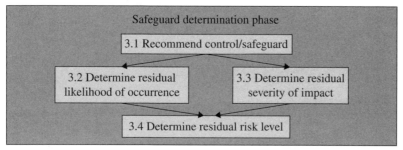

Figure 8.13   Risk assessment process flow.

Security risk analysis report   Once the security risks had been analyzed according to the severity and intensity thereof, an initial report must be compiled per assessed risk, listing all system components and establishing the system boundaries for the purpose of the report. System policies and procedures related to the risk must also be taken into consideration, when drafting the report (in order to define the risk and the required management).

The report must clearly state the list of identified threats and the related vulnerabilities, as well as the severity of the impact it may have and the likelihood of occurrence. This must go hand-in-hand with a list of suggested safeguards for controlling these threats and vulnerabilities. A list of recommended changes, with the appropriate levels of effort for each recommendation, must further be included in the report. Each suggested change must include the resulting reduction in risk, which will have to be achieved when implemented.

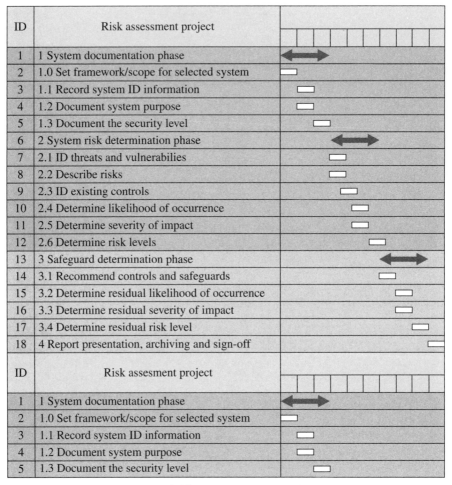

Figure 8.14   Risk assessment project.

TABLE 8.2   Severity of impact and risk levels

Insignificant. Will have almost no impact if threat is realized and exploits vulnerability.

Minor. Will have some minor effect on the system. It will require minimal effort to repair or reconfigure the system.

Significant. Will result in some tangible harm, albeit negligible and perhaps only noted by a few individuals or agencies. May cause political embarrassment. Will require some expenditure of resources to repair.

Damaging. May cause damage to the reputation of system management, and/or notable loss of confidence in the system's resources or services. It will require expenditure of significant resources to repair.

Serious. May cause considerable system outage, and/or loss of connected customers or business confidence. May result in compromise or large amount of Government information or services.

Critical. May cause system extended outage or to be permanently closed, causing operations to resume in a hot site environment. May result in complete compromise of Government agencies' information or services.

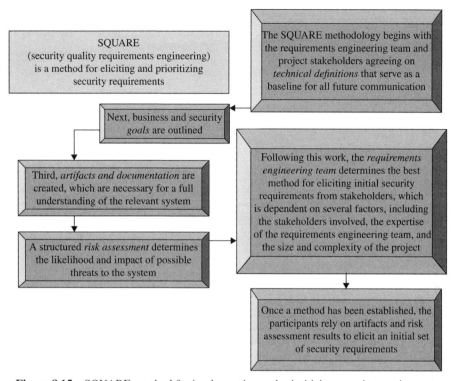

Figure 8.15   SQUARE: method for implementing and prioritizing security requirements.

Finally, the report must indicate the level of residual risk that would remain once the recommended changes are implemented.

SQUARE: Prioritizing security requirements

SQUARE is the abbreviation for Security Quality Requirements Engineering. It is an extremely valuable model assisting during the eliciting and prioritizing of security requirements. It starts with the technical definitions, serving as the baseline for all future communications between the requirements engineering team and project stakeholders.

This is followed by clear security goals, documenting the understanding of the relevant security system and the risk assessment, clearly defining all possible likelihoods and impacts. The best methods for eliciting the initial security requirements are drafted by the engineering team according to the size and complexity of the project. Finally, an initial set of security requirements are established based on risk assessment results and artifacts. Figure 8.15 depicts the earlier format.

The following steps are followed to make use of SQUARE, indicating the input and techniques required, the participants, and the desired outcome (Table 8.3).

TABLE 8.3   Steps for the use of SQUARE[a]

Steps to performing SQUARE		
Step 1: Agree on definitions	**Step 4: Perform risk assessment**	**Step 7: Categorize requirements as to level (system software etc.) and whether they are requirements or types of constraints**
Input: Candidate definitions from IEEE and other standards agencies	**Input:** Misuse cases, scenarios, security	**Input:** Initial requirements, architecture
Technique: Structured interviews	**Techniques:** Risk assessment method, analysis of anticipated risk against organizational risk tolerance, included threat analysis	**Techniques:** Work sessions using a standard set of categories
Participants: Stakeholders	**Participants:** Requirements engineer, risk expert, stakeholders	**Participants:** Requirements engineer, other specialists as needed
Output: Agreed to definitions	**Output:** Risk assessment results	**Output:** Categorized requirements
Step 2: Identify security goals	**Step 5 Select elicitation techniques**	**Step 8 Prioritize requirements**
Input: Definitions, candidate goals, business drivers, policies, procedures, examples	**Input:** Goals, definitions, candidate techniques, expertise of stakeholders, organizational style, culture, level of security needed, cost/benefit analysis, etc.	**Input:** Categorized requirements and risk assessment results
Technique: Facilitated work session, surveys, interviews	**Techniques:** Work session	**Techniques:** Prioritization methods such as triage, win–win, etc.
Participants: Stakeholders, requirements engineer	**Participants:** Requirements engineer	**Participants:** Stakeholders facilitated by requirements engineer
Output: Goals	**Output:** Selected elicitation techniques	**Output:** Prioritized requirements
Step 3: Artifacts to support security requirements	**Step 6: Elicit security requirements**	**Step 9: Requirements inspection**
Input: Potential artifacts (e.g., scenarios, templates, forms, etc.)	**Input:** Artifacts, risk assessment results, selected techniques	**Input:** Prioritized requirements, candidate formal inspection techniques

TABLE 8.3   *(Continued)*

	Steps to performing SQUARE	
Technique: Work sessions	**Techniques:** Joint application development, interviews, surveys, model based analyses, checklists, lists of reusable requirements types, document reviews	**Techniques:** Inspection method such as Fagan, peer reviews
Participants: Requirements engineer	**Participants:** Stakeholders facilitated by requirements engineer	**Participants:** Inspection team
Output: Needed artifacts, scenarios, models, etc.	**Output:** Initial cut at security requirements	**Output:** Initial selected requirements, documents of decision which record process and rationale

[a]Modelled after Mead NR, Viswanathan V, Padmanabhan D, Raveendran A. Incorporating security quality. Requirements Engineering (SQUARE) into Standard Life-Cycle Models. (CMU/SEI-2008-TN-006). Software Engineering Institute, Carnegie Mellon University, May 2008. http://www.sei.cmu.edu/publications/documents/08. reports/08tn006.html.

Security monitoring and enforcement

It is the responsibility of application coordinators to implement appropriate measures to detect attempts to compromise the security or integrity of information or information technology systems. When implementing monitoring capabilities, consideration should be given as to what situations are to be monitored based on the extent of risk, the most effective means for monitoring security activities, the resources available for monitoring, and system constraints that limit the ability to monitor security events. If appropriate measures are not available within a system environment to effectively monitor security events, additional controls should be implemented to mitigate security risks.

When activity occurs that is in conflict with security policies and standards, application coordinators should take the appropriate steps to enforce desired security practices. The steps involved range from training of the users, revoking access, altering security parameters, and possibly disciplinary actions.

The facts surrounding an intrusion or system compromise must be documented, reported to the security officer, and include the circumstances that led to the discovery of the incident, actions that were immediately taken, the names of persons involved in investigating the incident, and detailed observations about what transpired, what damage was caused, and what systems or files were compromised.

The security officer must enforce and support the security policy by responding to business ethics violations through disciplinary action, termination of services, suspension, or prosecution.

Security awareness program

It is the responsibility of management to ensure that all employees understand how to protect company assets, including information and information resources and comply with security policies, standards, and procedures. Supervisors and managers must ensure that persons working within their department understand general security requirements and that they are sufficiently knowledgeable about the security policies, standards, and procedures to recognize the need for protection and the requirements for which they are specifically responsible.

The security officer with assistance from the security team is responsible for developing and implementing an information security awareness program that supports employee awareness.

Managers and supervisors need to be aware of performance in this area, encourage good security practices, and address inappropriate behavior. Application coordinators can assist in implementing specific awareness programs.

Proposed future training requirements

The following are critical training requirements for security staff within the company. Application and levels will be determined by the appointments and responsibilities of staff:

- Semi-quantitative risk assessment techniques—Machinery based
- Hazards identification and analysis techniques
- Techniques for hazard identification and analysis—HAZOP
- Failure modes and effects analysis or "FMEA"
- Analysis of the consequences—mechanics of fire, explosion, and toxic releases
- Role of fault tree analysis to identify how accidents can happen
- Application to critical activities onshore and offshore—HAZOP
- The role of event tree analysis in scenario development.
- The role of fault tree analysis for multi-causation analysis
- Applications for ETA and FTA
- Human contribution to accidents
- The role of root cause analysis in identifying management system failures
- Accident investigation techniques I: fault tree analysis or "FTA"

Security management

The security manager has a different focus and responsibility than the rest of the organization. The rest of the organization is focused on providing production, research, shipping, and so on. The security manager is focused on avoiding losses through internal and external undefined sources.

Security department cuts across various disciplines. The security manager is more than a glorified guard force manager, although that is the general perception. The security manager cannot do the entire job alone; he or she needs subordinates, and the subordinates are the first line of defense and form the perception of the company by visitors. Sometimes, loss prevention and control is considered a part of security.

There is often a conflict between the organization and the people due to the differing histories of the people and their talents and abilities. This sometimes causes conflicts within the organization. Every organization has a formal and an informal organization chart for effectiveness—You know what the organization chart says, but when you want to get something done, who do you really go to?

Every organization has a culture, and it is often based on who has the ear of whom in management and what special privileges does that individual get or get away with, that is, the company doctor's parking space. It is important to recognize these factors and deal with them on a practical level.

The differing roles of the security department

The following are the different roles within the security department:

- Arrests and prosecutes persons committing attacks, theft, and so on
- Designs and implements physical control for access
- Conducts pre-employment screenings
- Monitors pertinent security information from military, police
- Administers vehicle parking
- Administers company lock and key policy
- May provide supervision of fire/rescue and medical services
- Conducts security indoctrination and training
- Investigates all criminal activity committed on company property
- Protects executives against kidnapping and extortion
- Conducts financial stability or due diligence on vendors, merger candidates, and so on
- Coordinates special protection during periods of civil unrest or disaster
- Contracts for outside security services as required

Stress management techniques

Everyone is subject to some stress, because it is a part of everyday life. Security personnel may have a bit higher stress than some other occupations because of the nature of their differing and sometimes contradictory roles in keeping the plant, its equipment, and personnel safe from various known and unknown hazards. Oftentimes, the security force must worry about the attacks from outside by persons unknown, with unknown armaments, plus worry about employee theft (and theft is not confined to plant workers, but it can also involve top management). Here are some suggestions for handling job stress.

Stress is another name for fear, and or worry. When we are worried or stressed, we cannot perform at our peak. Stress has physiological effects that include the following:

- Increased adrenaline in the blood
- Headaches
- High blood pressure
- May cause diabetes and other stress-related diseases
- Can cause heart attacks

The following techniques will help us to reduce or eliminate worry:

- Live in "day to day" with respect to worry.
- You have enough to worry about each day.
- Next week's deadlines are next week's worries.
- You can and must organize for upcoming events, but you cannot worry about things beyond your control.
- Worry about today's tasks today and do your best to achieve the end results.

How to analyze worry?

- Get all the facts—you cannot worry about what is unknown.
- Think about how you will approach the problem making you worry.
  - ◦ Try and develop two or three ways in which the problem can be solved or be made to go away.
  - ◦ Outline these scenarios and write them down.
  - ◦ Select the best solution.
- Weigh all the facts and then come to a decision.
- When you reach the decision—act.
- Save your thoughts and analysis in case you need it later.

How to break the habit of worrying?

- Keep busy an idle mind or hand is the devil's tool especially for worry.
- Worries can be significant or insignificant. Decide which worries are worth your time.
- Analyze what is making you worry.
- Do not worry about insignificant things. "Don't sweat the small stuff!"
- You may have to make a list of priority things for the day, but then review it for significance and re-prioritize so that you have the most significant items handled.
- On average, at least half or more of what we worry about never comes to pass.
- You have enough to worry about, and what are the chances that things will arrive or not.

- Chances are that the things you worry about may not come to pass—especially if you are going to be working to prevent them happening.
- Learn that there are some things we just cannot influence, and so do not worry about them.[1]
- You can learn from mistakes, but you cannot necessarily fix them. So what is done is done, and you need to move on and not worry about the mistakes. They are now in your past.
- Do not worry about the past. You have done your best, and it is not worthwhile trying to worry about what you did or did not do. You did your best at the time.

How to eliminate the causes of worry?

- Realize that listening to much of the news media will create unnecessary fear and worry. It is their job to get you worked up about things, and they do it well. So be skeptical and analytical when you listen to or read the news.
- Plant rumors are just like the news. Do not pay attention to rumors or gossip because they may be unfounded.
- Cultivate a positive mental attitude.
- PUSH—Pray until something happens.
- Fill your mind with thoughts of peace, courage, health and hope, religion, prayer, and positive and creative thoughts.
- Never consider revenge against your enemies, it is an unfruitful exercise.
- Expect ingratitude—especially in the guard force.
- Count your blessings not your troubles.
- Be polite and respectful to everyone—Do not try to imitate others.
- Create happiness for others—do something nice and unexpected for others for no good reason.
- Give unexpected appreciation to someone else.
- Try to rest before you get tired. You know that you are approaching tired, so take a few minutes to rest and refresh.
- Learn to relax at home.
- Apply good working habits by the following:
  - Clean desks help you organize and prioritize your work, by eliminating distractions.
  - List the things you have to do. Then prioritize them based on their importance and schedule.
  - Face a problem and solve it and make the decisions you need at the time.
  - Learn to organize and delegate.
  - Be enthusiastic about your work.

Security management techniques

Theory X of human relations management Often based on **autocratic styles**, it assumes the worst about employees. Average human dislikes work and will avoid it if possible. Because of dislike of work, most people must be coerced, controlled, directed, threatened, and so on, to get adequate performance from them. Average human prefers to be directed, wishes to avoid responsibility, has little ambition, and wants security above all else.

Theory Y management style The average individual considers work as a part of life, and as natural as play or rest. External control and the threat of punishment are not the only ways of bringing about effort. Man will exercise self-direction and self-control when he has committed himself to the work. Rewards (part) are found in the execution and satisfaction of the work. Average person seeks responsibility under the right conditions—it gives him a sense of pride. People can be imaginative and creative in the fulfillment of their work if given a chance. The intellectual potential of an individual is only partially utilized by his work. The challenge is to get the commitment to utilize that intellectual role to its fullest.

Maintenance factors (job surroundings) The following maintenance factors will have an influence on the behavior of employees:

- Pay
- Status
- Policy and administration
- Interpersonal relationships
- Benefits
- Supervision
- Working conditions
- Job security

Motivators (the job itself) The following aspects can be regarded as motivators within the job environment and should be manage well. It must further be understood clearly that each individual has different personalities, backgrounds, upbringing, and skills. Even levels of maturity and experience will differ from person to person:

- Responsibility
- Achievement
- Recognition
- Advancement
- Growth

Bad management traits Bad management traits are serious and detrimental to any form of management. Management within the security domain is even more adversely affected by these forms of managerial conduct:

- **Nepotism rather than quality.** Is the manager's relative given an unwarranted promotion, or is the manager looking out for his kinfolk or clan? Unfortunately, in many societies, particularly in the Middle East, this tends to happen with unfortunate regularity.

- **Playing favorites.** Does the manager give "good" assignments to a select few, or are those people really that extra qualified?

- **Manager who wants to be liked rather than respected.** This is a common trait in supervisors and middle managers, and others who move up the ranks.

- **Difference between leadership and management.** The acknowledged difference between leadership and management is characterized by the following: Management says, "Here's what I want you to do." Leadership says, "Let us to the following…." One is participative, the other is directive.

- **Being "one of the guys" can create avoidance of unpopular decisions that must be made.** No manager likes to make unpopular decisions, but events happen and decisions have to be made.

- **Fine line between being respected and being feared!** Many managers cross this line. When female employees are in the workforce, they should also be treated with respect and not bullied. In masculine cultures, this is often a difficult challenge.

- **Ivory tower "know-it-all" manager.** This guy deserves to fail. The adage here is to remind people that, "The quickest way to foul up a project is to tell an individual to do, 'Exactly as I tell you.' Because it alleviates the individual from using judgment and responsibility."

- **Improper delegation.** This is a classic and common mistake. If you are going to delegate responsibility, delegate the authority to accomplish the task. Otherwise, It is like telling the worker to drive downtown, but not providing him with a car.

- **Blind insistence on the "company way."** There is always more than one way to complete a project, and new and different is not necessarily wrong.

- **Manager who ignores need for others and his own growth.** This is more than an issue of training. People like to grow in their work and handle new and different challenges that can stretch them. It prevents them from being bored.

- **Manager who fails to give proper credit to subordinates.** This guy is an egotistical thief who is only looking for his own aggrandizement. Make sure that is not you by giving proper credit where it is due.

- **RHIP Manager—Elitism.** RHIP is short for Rank Hath Its Privilege. It is one thing for the manager to have a better car, or lifestyle because of earned salary, but when special favors are put in the job because of rank, well, that is just wrong!

- **Excessive secretiveness—failure to share information.** The security business is often about secrecy and the "need to know." Make sure that you share critical decisions with your subordinates.

- **Manager who views disciplinary process as punitive.** This is one of the worst sins of management. Discipline' root is the word "Disciple." The best idea is to have a corrective process which points out the errors with a way forward so that the person receiving the discipline is led to the right path and shown the error of his or her ways and how to correct them.
- **Manager who is unreliable and/or two faced.** Worst type of manager ever!
- **Manager who avoids decisions.** A lack of decision on an issue is really a decision—a decision NOT to decide. Avoiding an issue would not make it go away.
  - The failure to decide is in itself a decision. See the preceding item.
  - The consequences of inaction are almost always worse than the wrong decision.
- **The manager who is a slave driver.**
- **The crisis manager.**
  - If everything is a crisis, there is no set of priorities.
  - Your failure to plan is not cause for me to have a crisis.
  - Crises can only happen occasionally.

CONCLUSION

Security is everyone's business. We need to approach it in a professional manner with intelligence and personnel training on the important things. The people in the security force are professionals and deserve respect just as much as the plant operators and engineers. In that regard, we hope this book has been informative and helpful.

NOTES

1 As an example: You have heard that the plant is going to have layoffs, but nothing more specific than that. So, while it is hard to not worry about keeping your job, there is little that you can do once a decision is made. It is not fatalism, but realism.

  You can plan to polish up your CV, and maybe even circulate it to other companies if you feel that your job is in jeopardy, but you do not know, until you know. And, premature action often takes one from the frying pan into the fire. This has happened to us under a number of difference circumstances.

2 Trojan Powder Policies clearly stated that, "horseplay or running without cause is immediate cause for dismissal, regardless of the level of the employee." The rationale for this is because in a Powder Works where things can go boom very quickly, the sight of someone running would cause anyone to assume that the individual is running to safety and they would naturally drop whatever they were doing and run in the same direction.

PHYSICAL SECURITY CHECKLIST

BUILDING

1. Facility Address:

2. Description and number of buildings:

3. Compounds/products manufactured, tonnages (attach list)

Industrial Security: Managing Security in the 21st Century, First Edition. David L. Russell and Pieter C. Arlow.
© 2015 John Wiley & Sons, Inc. Published 2015 by John Wiley & Sons, Inc.

4. Facility hours/office hours/shift times

	Hours	Personnel	No. of supervisors
M-F office			
Weekend/holiday office			
Shift hours			
Weekdays			
Weekends/holidays			

5. Attach detailed plant maps showing production areas, fence lines, and buildings or annotate an aerial photograph.

6. Photograph and show location of gates and entrances. Label and attach one picture per page.

7. Location of primary and secondary security centers

8. Is all information channeled into both centers from all sensors/cameras, fences/ gates, doors, etc. Independently?

9. Size and number of guards/security force personnel

	Hours	Personnel	No. of supervisors
M-F office			
Weekend/holiday office			
Shift hours			
Weekdays			
Weekends/holidays			

10. Fence line security: Describe fences and sensors. Attach a drawing, if necessary.

11. Is there a distance between vegetation and fence line?_____

12. Are property boundaries clearly identified? _____

13. Is fence line illuminated at night? _____

14. Are the gates in good condition? _____

15. Are the gate hinges internal or external? _____

16. How are unused gates protected and monitored? _____

_____

_____

17. Are there dark spots in the fencing illumination? _____

18. Is the fence line monitored by closed circuit TV? _____

19. Color TV or black and white? _____

20. Is the TV equipped with infrared sensors? _____

21. Is there at least 10 m between fence line and shrubbery and trees and poles _____

22. Is the boundary fencing clearly marked with "No Trespassing" signs? _____

23. Are all drains and swales larger than 40 cm screened for antipersonnel intrusion? _____

24. Is fencing continuous with all rises and dips in the terrain?

25. Is the fence line illumination automatic? _____

26. Is there a manual override for illumination controlled by the guard station or central dispatch station? _____

27. If there are holes in the fence, what is the plan and schedule for making repairs?

28. Are all buildings that are not continually occupied locked after business/shift hours?

29. Are all building doors and windows facing fencing locked, and do they have alarms? _____

30. Are the alarms internal and external? _____

31. Are there CCTV on exterior doors? _____

_____

32. Do these alarm register on the central dispatch station? _____

33. What is the response time for inspection of a gate or door alarm? _____

34. List and describe building doors and ground floor windows and their alarms and cameras.

35. Are all buildings that are not continually occupied locked after business/shift hours?

LOCK AND KEY, ALARM SYSTEMS, AND GUARDS

36. Are keys signed for?

37. Who controls duplicate keys?

38. Is there a master list for keys and their access?

39. Who keeps the key logs?

40. List the number of keys in each area/building and who controls them on a separate sheet.

41. Are all the keys accounted for?

42. When an employee leaves, do they surrender their badge and keys?

43. Are keys nonstandard so that they cannot be easily duplicated?

44. Are padlocks on a separate system with separate keys?

45. Describe the following about the building, gate, and fence alarm system.

	Doors	Fence line	Gates
Manufacturer			
Type			
Installation date			
Inspection dates			
Serviced by:			

46. How frequently are alarm systems tested? _____

47. Date of last test? _____

48. Where does alarm system terminate? _____

49. Does each alarm generate a printed report? _____

50. Who keeps the records? _____

51. Do the alarms show a pattern? _____

52. Where is the alarm panel located? _____

53. Is the alarm panel located behind locked doors? _____

54. Who has the authority to turn alarms on and off? _____

55. Does the alarm system have a battery backup? _____

56. Is the battery backup periodically checked and maintained? _____

57. How frequently? _____

58. Do the guard patrols have duress alarms? _____

59. How frequently are the guard forces given polygraph tests? _____

60. Do the guards know CPR? _____

61. Do the guards maintain fitness levels? _____

62. What are the physical standards for the guard force? _____

63. Distance between fence lines and closest housing/industrial/commercial units not owned by company on each compass direction?

N	
S	
E	
W	

64. Attach a wind rose for history of prevailing winds.

65. Does plant have any remote operations? Yes _____ No _____

66. If Yes, describe how remote operations are monitored.

67. What is posted speed approaching gates/entrances? _____

68. What barriers exist to prevent vehicles from crashing gates/entrances?

69. Where are inspections conducted on all incoming vehicles? Describe.

70. Briefly describe how shipments by rail are inspected and allowed into the plant. Attach a separate description, if required.

71. Is shipping and receiving located remotely from the balance of the plant operations? If yes, how far? _____

72. Is there a procedure for insuring that drivers of vehicles cannot enter the plant? Describe and determine if barriers can stop an incoming moving vehicle

73. Are the security force armed? _____

74. How frequently do they undergo firearms accuracy training? _____

75. How are detentions or arrests handled?

76. Does the night shift/graveyard shift have regular rounds?

77. Does the night shift/graveyard shift have key stations?

78. Do all the security force have encoded radios? Can they monitor in-plant communications?

79. Does the security force have a way of communicating with plant fire department? _____ Off-site Security (Police and Fire) _____ Hospital _____

EMPLOYEE SECURITY

80. Is employee ingress and egress restricted to controlled and monitored locations? _____

81. Does security have a way to know how many employees are in the plant at any one time, and what locations they are supposed to be working in? _____

82. Employee/visitor/contractor's ingress/egress is controlled by?
    Badge _____
    Pass _____
    Guard _____
    Key _____
    Receptionist _____

83. Does the plant use contractor labor for maintenance? _____

84. Have all contractor laborers been screened? _____

85. Do all employees have badges? _____

86. Do all contractors/visitors have different badges than employees? _____

87. List companies and names, and contact information for all contractors on a separate sheet.

88. Are visitors escorted? _____

89. Are personal electronics/laptops/flash drives, disks, etc., allowed in plant? _____

90. Are bags and personal possessions searched by everyone entering and leaving the plant?

91. Do the ingress and egress locations from the plant have metal detectors?

92. Do the employees and visitors have a designated parking area outside the plant?

93. Is the parking area secure and illuminated? _____ CCTV monitored? _____

TRASH REMOVAL AND SHIPMENTS FROM THE FACILITY

94. Are outgoing shipments inspected and manifests verified?_____

    Describe the procedure. _____

95. Name and address of trash removal service._____

96. How often is trash removed? _____

97. Are trash removals supervised by security force? _____

98. How often is trash inspected? _____

99. Are trash and hazardous wastes hauled by the same company?
    _____

100. Where are hazardous wastes disposed of?_____

PLANNING

101. Does the plant have plans for the following, and are they up to date:
     a. Plantwide emergency
     b. Evacuation
     c. Bomb
     d. Fire
     e. Extreme weather (Tornado and/or Hurricane)
     f. Flood
     g. Earthquake
     h. Explosion
     i. Loss of utility services
     j. Civil disorder
     k. Terrorist attack
     l. Spill events
102. List the last time each plan was tested?

103. Does the plant follow Incident Command System

104. Do each of the plans have the contact information for all necessary key employees?

105. Are the evacuation routes and assembly points posted?

106. Is there a plant map showing the location of fire extinguishers?

107. Is there a plant map showing the location of spill control materials?

108. Is there coordination with the Local Emergency Planning Committee?

109. Have any or all of these plans been submitted/shared with the local fire and police forces and local hospital to provide services?

110. Are the routes to the local hospital posted and available?

MAIL HANDLING

111. How is the mail handled?
   a. Incoming
   b. Outgoing
   c. Are there procedures for hazardous materials/bombs in the mail?
   d. Are all mails opened?

FIRE PLANS

112. Are all buildings up to fire code?_____

113. Has the fire marshal been into inspect the buildings? _____

114. Is the plant equipment and process conforming to current fire codes?
   _____

115. Is the fire sprinkler system tested regularly?_____

PERSONNEL SECURITY	YES	NO
1. Does your staff wear ID badges?		
2. Is a current picture part of the ID badge?		
3. Are authorized access levels and type (employee, contractor, visitor) identified on the badge?		
4. Do you check the credentials of external contractors?		
5. Do you have policies addressing background checks for employees and contractors?		
6. Do you have a process for effectively cutting off access to facilities and information systems when an employee/contractor terminates employment?		
PHYSICAL SECURITY		
7. Do you have policies and procedures that address allowing authorized and limiting unauthorized physical access to electronic information systems and the facilities in which they are housed?		
8. Do your policies and procedures specify the methods used to control physical access to your secure areas, such as door locks, access control systems, security officers, or video monitoring?		
9. Is access to your computing area controlled (single point, reception or security desk, sign-in/sign-out log, temporary/visitor badges)?		
10. Are visitors escorted into and out of controlled areas?		
11. Are your PCs inaccessible to unauthorized users (e.g., located away from public areas)?		
12. Is your computing area and equipment physically secured?		
13. Are there procedures in place to prevent computers from being left in a logged on state, however briefly?		
14. Are screens automatically locked after 10 minutes idle?		

Industrial Security: Managing Security in the 21st Century, First Edition. David L. Russell and Pieter C. Arlow.
© 2015 John Wiley & Sons, Inc. Published 2015 by John Wiley & Sons, Inc.

15. Are modems set to Auto-Answer OFF (not to accept incoming calls)?
16. Do you have procedures for protecting data during equipment repairs?
17. Do you have policies covering laptop security (e.g., cable lock or secure storage)?
18. Do you have an emergency evacuation plan and is it current?
19. Does your plan identify areas and facilities that need to be sealed off immediately in case of an emergency?
20. Are key personnel aware of which areas and facilities need to be sealed off and how?

ACCOUNT AND PASSWORD MANAGEMENT

21. Do you have policies and standards covering electronic authentication, authorization, and access control of personnel and resources to your information systems, applications and data?
22. Do you ensure that only authorized personnel have access to your computers?
23. Do you require and enforce appropriate passwords?
24. Are your passwords secure (not easy to guess, regularly changed, no use of temporary or default passwords)?
25. Are your computers set up so others cannot view staff entering passwords?

CONFIDENTIALITY OF SENSITIVE DATA

26. Do you classify your data, identifying sensitive data versus non sensitive?
27. Are you exercising responsibilities to protect sensitive data under your control?
28. Is the most valuable or sensitive data encrypted?
29. Do you have a policy for identifying the retention of information (both hard and soft copies)?
30. Do you have procedures in place to deal with credit card information?
31. Do you have procedures covering the management of personal private information?
32. Is there a process for creating retrievable back up and archival copies of critical information?
33. Do you have procedures for disposing of waste material?
34. Is waste paper binned or shredded?
35. Is your shred bin locked at all times?

36. Do your policies for disposing of old computer equipment protect against loss of data (e.g., by reading old disks and hard drives)?
37. Do your disposal procedures identify appropriate technologies and methods for making hardware and electronic media unusable and inaccessible (such as shredding CDs and DVDs, electronically wiping drives, burning tapes, etc.)?

DISASTER RECOVERY

38. Do you have a current business continuity plan?
39. Is there a process for creating retrievable back up and archival copies of critical information?
40. Do you have an emergency/incident management communications plan?
41. Do you have a procedure for notifying authorities in the case of a disaster or security incident?
42. Does your procedure identify who should be contacted, including contact information?
43. Is the contact information sorted and identified by incident type?
44. Does your procedure identify who should make the contacts?
45. Have you identified who will speak to the press/public in the case of an emergency or an incident?
46. Does your communications plan cover internal communications with your employees and their families?
47. Can emergency procedures be appropriately implemented, as needed, by those responsible?

SECURITY AWARENESS AND EDUCATION

48. Are you providing information about computer security to your staff?
49. Do you provide training on a regular recurring basis?
50. Are employees taught to be alert to possible security breaches?
51. Are your employees taught about keeping their passwords secure?
52. Are your employees able to identify and protect classified data, including paper documents, removable media, and electronic documents?
53. Does your awareness and education plan teach proper methods for managing credit card data (PCI standards) and personal private information (Social security numbers, names, addresses, phone numbers, etc.)?

COMPLIANCE AND AUDIT		
54. Do you review and revise your security documents, such as: policies, standards, procedures, and guidelines, on a regular basis?		
55. Do you audit your processes and procedures for compliance with established policies and standards?		
56. Do you test your disaster plans on a regular basis?		
57. Does management regularly review lists of individuals with physical access to sensitive facilities or electronic access to information systems?		
Checklist Response Analysis For each question that is marked "No," carefully review its applicability to your organization. Implementing or improving controls decreases potential exposure to threats/vulnerabilities that may seriously impact the ability to successfully operate.		

CYBER SECURITY THREAT/VULNERABILITY ASSESSMENT

Impact scale likelihood scale

For this assessment, numeric rating scales are used to establish impact potential (0–6) and likelihood probability (0–5).

IMPACT SCALE	LIKELIHOOD SCALE
1. Impact is negligible	0. Unlikely to occur
2. Effect is minor, major agency operations are not affected	1. Likely to occur less than once per year
3. Organization operations are unavailable for a certain amount of time, costs are incurred. Public/customer confidence is minimally affected	2. Likely to occur once per year
4. Significant loss of operations, significant impact on employee/user confidence	3. Likely to occur once per month
5. Effect is disastrous, systems are down for an extended period of time, systems need to be rebuilt and data replaced	4. Likely to occur once per week
6. Effect is catastrophic, critical systems are offline for an extended period; data are lost or irreparably corrupted; employees safety, and possibly external public are affected	5. Likely to occur daily

When determining impact, consider the value of the resources at risk, both in terms of inherent (replacement) value and the importance of the resources (criticality) to the organization's successful operation.

Factors influencing likelihood include: threat capability, frequency of threat occurrence, and effectiveness of current countermeasures (security controls). Threats caused by humans are capable of significantly impairing the ability for an organization to operate effectively.

Human threats sources include:

SOURCE	SOURCE DESCRIPTION
Insiders:	Employees,
General contractors	Cleaning crew, developers, technical support person-
and subcontractors	nel, and computer and telephone repair workers
Former employees	Includes those who have quit, were fired, or retired
Unauthorized personnel	Intruders, computer hackers, trespassers, terrorists,
	and other people with bad intent

Score risk level risk occurrence result

21–30 High Risk Occurrence may result in significant loss of major tangible assets, information, or information resources. May significantly disrupt the organization's operations or seriously harm its reputation.

11–20 Medium Risk Occurrence may result in some loss of tangible assets, information, or information resources. May disrupt or harm the organization's operation or reputation. For example, authorized users are not able to access supportive data for several days.

1–10 Low Risk Occurrence may result in minimal loss of tangible assets, information, or information resources. May adversely affect the organization's operation or reputation. For example, authorized users are not granted access to supportive data for an hour.

HUMAN THREATS	Impact (0–6)	Probability (0–5)	Score (Impact × Probability)
1. Human error			
• Accidental destruction, modification, disclosure, or incorrect classification of information	____	____	____
• Ignorance: inadequate security awareness, lack of security guidelines, lack of proper documentation, lack of knowledge	____	____	____
• Workload: Too many or too few system administrators, highly pressured users	____	____	____

HUMAN THREATS	Impact (0–6)	Probability (0–5)	Score (Impact × Probability)
• Users may inadvertently give information on security weaknesses to attackers	___	___	___
• Incorrect system configuration	___	___	___
• Security policy not adequate	___	___	___
• Security policy not enforced	___	___	___
• Security analysis may have omitted something important or be wrong	___	___	___
2. Dishonesty: Fraud, theft, embezzlement, selling of confidential agency information	___	___	___
3. Attacks by "social engineering"			
• Attackers may use telephone to impersonate employees to persuade users/administrators to give user name/passwords/modem numbers, etc.	___	___	___
• Attackers may persuade users to execute Trojan Horse programs	___	___	___
4. Abuse of privileges/trust	___	___	___

GENERAL THREATS	Impact (0–6)	Probability (0–5)	Score (Impact × Probability)
1. Unauthorized use of "open" computers/laptops'	___	___	___
2. Mixing of test and production data or environments	___	___	___
3. Introduction of unauthorized software or hardware	___	___	___
4. Time bombs: Software programmed to damage a system on a certain date	___	___	___
5. Operating system design errors: Certain systems were not designed to be highly secure	___	___	___
6. Protocol Design Errors: Certain Protocols not designed to be highly secure. Protocol weakness in TCP/IP can result in:	___	___	___
• Source Routing, DNS Spoofing, TCP sequence guessing, unauthorized access	___	___	___

HUMAN THREATS	Impact (0–6)	Probability (0–5)	Score (Impact × Probability)
• Hijacked sessions and authentication session/transaction replay, data is changed or copied during transmission	____	____	____
• Denial of Service due to ICMP Bombing, TCP-SYN Flooding, large PING packets, etc.	____	____	____
7. Logic Bombs: Software programmed to damage a system under certain conditions	____	____	____
8. Viruses in programs, documents and e-mail attachments	____	____	____
9. Trojan Horses (programs masquerading as other programs)	____	____	____
10. Intruders seeking to spoof or obtain unauthorized access	____	____	____
11. Phishing attacks (e-mail appearing to come from authorized sources)	____	____	____
12. Attacks through SKYPE®, VIBER®, and teleconferencing software	____	____	____
13. SCADA attacks and control system spoofing	____	____	____
14. Sabotage	____	____	____
15. Physical destruction of equipment	____	____	____
16. Electromagnetic radiation (EMP) attacks	____	____	____
17. Failure to properly remove and destroy electronic media in abandoned equipment	____	____	____
18. Deliberated deletion of critical files and backup systems	____	____	____

CYBER SECURITY THREAT/VULNERABILITY ASSESSMENT SCORING

Next steps

After completing a review of current security controls and along with a review and rating of potential threats/vulnerabilities, a series of actions should be determined to reduce risk (threats exploiting vulnerabilities) to and acceptable level. These actions should include putting into place missing security controls, and/or increasing the strength of existing controls.

Security controls should ideally reduce and/or eliminate vulnerabilities and meet the needs of the business. Cost must be balanced against expected security benefit and risk reduction. Security remediation efforts and actions will be focused on addressing identified high risk threat/vulnerabilities.

The recommended Security Actions to remediate vulnerabilities should be displayed in a tabular form, and a color coded or shaded table displaying risk vulnerabilities, costs in a semi-quantitative format could be formulated to permit rapid identification of the risks accompanying the report. Another approach is to conduct a Bow-Tie Analysis of the various risks and preventative/remedial measures and cost the damages and remedial responses in the table. An example of the type of table is shown below.

Sample of Cyber Risk Vulnerability/Risk and Preventive Actions				
Item	Damage Cost, Millions	Probability of Occurrence, No. per year	Remedial Actions Equipment Costs, Millions	Training and Other Associated Costs, Millions
Spoofing SCADA—destroys process plants	2.0	0.01	0.5	0.0
Intrusion	0.04	1.0	0.02	0.01
Cyber attack	0.3	12.0	0.05	0.05
Data breach	0.3	1.0	0.01	0.05

The table is more effective in color, but even in one color, it can be printed with shading from the darkest to the lightest shading to highlight the important information. A variety of display options can be used.

INDEX

Industrial Security: Managing Security in the 21st Century, First Edition. David L. Russell and
Pieter C. Arlow.
© 2015 John Wiley & Sons, Inc. Published 2015 by John Wiley & Sons, Inc.